AF846590

Contributions to Management Science

For further volumes:
http://www.springer.com/series/1505

Nils Rüfenacht

Implicit Embedded Options in Life Insurance Contracts

A Market Consistent Valuation Framework

Physica-Verlag
A Springer Company

Nils Rüfenacht
Baumgartenweg 23
Basel
Switzerland

ISSN 1431-1941
ISBN 978-3-7908-2842-9 e-ISBN 978-3-7908-2843-6
DOI 10.1007/978-3-7908-2843-6
Springer Heidelberg Dordrecht London New York

Library of Congress Control Number: 2012934048

© Springer-Verlag Berlin Heidelberg 2012
This work is subject to copyright. All rights are reserved, whether the whole or part of the material is concerned, specifically the rights of translation, reprinting, reuse of illustrations, recitation, broadcasting, reproduction on microfilm or in any other way, and storage in data banks. Duplication of this publication or parts thereof is permitted only under the provisions of the German Copyright Law of September 9, 1965, in its current version, and permission for use must always be obtained from Springer. Violations are liable to prosecution under the German Copyright Law.
The use of general descriptive names, registered names, trademarks, etc. in this publication does not imply, even in the absence of a specific statement, that such names are exempt from the relevant protective laws and regulations and therefore free for general use.

Printed on acid-free paper

Physica-Verlag is a brand of Springer
Springer is part of Springer Science+Business Media (www.springer.com)

Implicit Embedded Options in Life Insurance Contracts.
A Market Consistent Valuation Framework

Dissertation

zur Erlangung der Würde eines Doktors der Staatswissenschaften

vorgelegt der

Wirtschaftswissenschaftlichen Fakultät der Universität Basel

von Nils Rüfenacht

von Meikirch (Bern)

Physica-Verlag
A Springer Company

Basel 2011

Genehmigt von der Wirtschaftswissenschaftlichen Fakultät der Universität Basel auf Antrag von Prof. Dr. Heinz Zimmermann und Prof. Dr. Herbert Lüthy.

Basel, den 29. August 2011

Der Dekan
Prof. Dr. George Sheldon

"If you can meet with triumph and disaster
and treat those two impostors just the same"

Extract from "If" by Rudyard Kipling

Acknowledgements

After having spent an extremely interesting and instructive time doing research with a couple of ups and downs, this thesis has finally come to an end.

I am deeply grateful to my supervisor, Prof. Dr. Heinz Zimmermann, who gave me the opportunity to realise this thesis and all his constructive critiques and suggestions throughout my PhD studies. The same thank goes to Prof. Dr. Herbert Lüthy, my co-supervisor.

Special thanks goes to my employer, Helvetia Versicherungen, that supported me with important data and enabled this thesis to become related to practice. Furthermore, I would like to express my thanks to my fellow workers Dr. Roman Reith, Vincent Rey and Patrick Stöckli for the fruitful discussions and interesting inputs from the practitioner's view.

A further thanks goes to Lea Lang who patiently helped me to correct my English spelling.

Finally, I say thank you very much to my beloved girlfriend Norina Martig for her moral support during the final stage of this thesis and to Lily Toberer, who often inspired me by her bright nature.

Abstract

During the eighties and the early nineties of the past century many life insurance companies added so called embedded options as an additional feature to their policies. Interest rate guarantee, surrender, dividend or guaranteed annuity option are popular examples for this. These options have in common that they are all very sensitive to interest rates.

Looking back on this period, one observes an astonishing high level of market interest rates over several years. Assuming interest rates to remain on that level there was no obvious need to deeply care about proper pricing or hedging of the embedded options when new policies were issued.

When interest rates started to fall from the mid-nineties many embedded options suddenly turned into the money. Due to the typical long term contract duration in life insurance business insurers more and more suffered under this growing financial burden. British Equitable Life Assurance Society that had to close to new business by the end of 2000 impressively demonstrated how serious the situation had become.

The collapse of Equitable Life as a result of its embedded options is, amongst others, one of the reasons that caused a far-reaching change in the local insurance supervisory and general accounting methodology taking place all over the world at the beginning of this century. Rule-based regulation was replaced by principles-based regulation enabling insurers to individually assess their own risk exposure in a constantly changing economic environment.

This thesis presents a market consistent valuation framework for implicit embedded options in life insurance contracts. This framework is used to perform an empirical analysis based on more than 110,000 real existing in-force life insurance policies and with a focus on the modelling of interest rates. Its results are the answer to the central question asked in the Objectives: *What value do the considered embedded options and guarantees have, absolutely and relative to the current policy reserves, given a perspective taken from the insurer, the policyholder and the shareholder respectively?*

As for the whole thesis, current is meant to be as at the end of 2009.

Within the empirical analysis interest rate guarantee, dividend and shareholder option are the considered embedded options. Below Table 1 summarises its values based on all the available data under best estimate assumptions for a representative set of surplus parameters. The relative values of the embedded options are obtained by dividing by the current policy reserve (including the current surplus reserve) P.

Table 1 Total values in Swiss Francs of the interest rate guarantee, the dividend option and the shareholder option under best estimate assumptions. The underlying data basis consists of 25,707 single premium policies and 85,730 regular premium policies with a current policy reserve (including the current surplus reserve) P of 1,157,664,264 Swiss Francs (single premium) and 1,642,879,688 Swiss Francs (regular premium)

	Single premium	Regular premium
Total interest rate guarantee	39,497,246	42,686,809
Interest rate guarantee per policy	1,536	498
Interest rate guarantee relative to P	0.0341	0.0260
Total dividend option	89,501,240	893,824,435
Dividend option per policy	3,482	10,426
Dividend option relative to P	0.0773	0.5441
Total shareholder option	18,358,262	102,467,939
Shareholder option per policy	714	1,195
Shareholder option relative to P	0.0159	0.0624
Total of all options	147,356,748	1,038,979,183
Total of all options per policy	5,732	12,119
Total of all options relative to P	0.1273	0.6324

The dividend and the shareholder option are implicitly hedged by definition already. Hence, the insurer does not need to set up a reserve for these options today. Its values basically quantify the worth additionally offered to the policyholders and the shareholders due to the currently applied surplus policy.

Different than surpluses the insurer guarantees crediting a fixed minimum interest rate on the savings premiums and this guarantee is neither implicitly nor explicitly hedged in this model. Thus, the value of the interest rate guarantee quantifies the amount the insurer should reserve today in order to meet this future obligations.

Stochastic ZCB prices are modelled as a part of the entire modelling process. Therefore, the impacts of three different interest rate models are considered: A single factor Cox-Ingersoll-Ross (CIR) model, a single factor Hull-White (HW) model and a two-additive-factor Gaussian (G2++) model.

The CIR model turns out to be unsuitable for the purpose of this modelling approach as it significantly underestimates the interest rate risk in terms of the low values of the interest rate guarantee. The G2++ has the edge over the HW model and produces more conservative results in terms of a higher total value of the interest rate guarantee, the dividend and the shareholder option. A summary on the total option values for the different interest rate models is highlighted in Table 2.

Table 2 The total option values in Swiss Francs for different interest rate models. The values in brackets denote the relative difference to the default model: the G2++ model

	Single premium	Regular premium
Total option value for the CIR model	109,907,929 (−0.2541)	860,577,698 (−0.1717)
Total option value for the HW model	137,484,798 (−0.0670)	1,024,063,971 (−0.0144)
Total option value for the G2++ model	147,356,748 (0.0000)	1,038,979,183 (0.0000)

At the same time different interest rate curves for discounting purposes are analysed. Both, Solvency II and the MCEV calculations, prescribe the use of the swap rate curve. An alternative comes in the form of the zero coupon curve, as it is required for the SST calculations. Using the current zero coupon rates reflects a more conservative behaviour in the interest rate assumptions but anyhow produces lower total values for the interest rate guarantee and higher total values for the dividend and the shareholder option. The total values of all the embedded options are summarised in Table 3.

Table 3 The total option values in Swiss Francs for different market interest rates. The values in brackets denote the relative difference to the interest rates used as default: the swap rates

	Single premium	Regular premium
Total option value using zero coupon rates	158,288,684 (0.0742)	1,128,444,234 (0.0861)
Total option value using swap rates	147,356,748 (0.0000)	1,038,979,183 (0.0000)

The most fundamental statement of this thesis is deduced from a comparison over time. Performing best estimate simulation runs as at the end of 2004 and 2009 impressively demonstrates where the limitations of the market consistent valuation methodology are when applying a modelling approach mixing both, market consistent and statutory valuation methodology. According to this, insurers should successively move towards a comprehensive and entirely consistent application of the market consistent methodologies regarding valuation, reserving and, as far as possible, also pricing. This is the corollary to get even more meaningful values for the embedded options in particular and to attain an entirely consistent risk management in general.

Contents

List of Figures

List of Tables

Acronyms

ALM	asset liability management
ANAV	adjusted net asset value
ATM	at the money
CAO	current annuity option
CEIOPS	Committee of European Insurance and Occupational Pensions Supervisors
EC	European Comission
EU	European Union
EV	Embedded Value
FINMA	Swiss Financial Market Supervisory Authority
FOPI	Federal Office of Private Insurance
GAO	guaranteed annuity option
ifp	in-force probability
MCEV	Market Consistent Embedded Value
MCR	minimum capital requirement
ML	maximum likelihood
P&L	pension and life insurance
QIS	qualitative impact studies
RBC	risk bearing capital
SAA	Swiss Association of Actuaries
SCR	solvency capital requirement
SDE	stochastic differential equation
SIA	Swiss Insurance Association
SPI	Swiss Performance Index ®
SST	Swiss Solvency Test
TC	target capital
ZCB	zero coupon bond

Chapter 1
Introduction

In the past 10 years the financial stability of many life insurers has been severely threatened by several sudden happenings but also by some structural changes. The burst of the dot-com bubble at the beginning of this century, the terrorist attacks on 11 September 2001 or the property crisis that finally led to a global financial crisis during 2008 and 2009, are only some selected examples of unexpected impacts that shocked capital markets worldwide including the insurance industry. In addition, I mention increasing longevity and lasting low market interest rates as slow-acting changes that caused further actuarial and financial challenges.

When I talk about low market interest rates, this is always a relative statement and I need to define my perspective. Analysing a period of the past 30 years shows that we had (relatively) high market interest rates all around the world from the beginning of the 1980s lasting approximately 10–12 years. Subsequent to this period, market interest rates continuously decreased and did not recover to that former level until now. In Fig. 1.1 the past evolution of market interest rates gets quantified. For comparative purposes I have selected the Swiss and UK 3 month interbank rate and the equivalent in the US market, the US 3 month certificate of deposit rate (CD rate).

Considering the balance sheet of a typical life insurer, we realise that a lion's share of its assets is invested in interest rate sensitive securities. Thus, it is self-explanatory that interest rates definitely become a relevant issue for a life insurance company. But this is only half of the story. Since middle of last century it has become more and more customary to give the policyholder additional benefits by embedding options and guarantees into their contracts. Minimum interest rate guarantee, surrender option or guaranteed annuity option are just some feasible but common examples thereof. And we also see that these options and guarantees are highly interest rate sensitive, too.

N. Rüfenacht, *Implicit Embedded Options in Life Insurance Contracts*,
Contributions to Management Science, DOI 10.1007/978-3-7908-2843-6_1,
© Springer-Verlag Berlin Heidelberg 2012

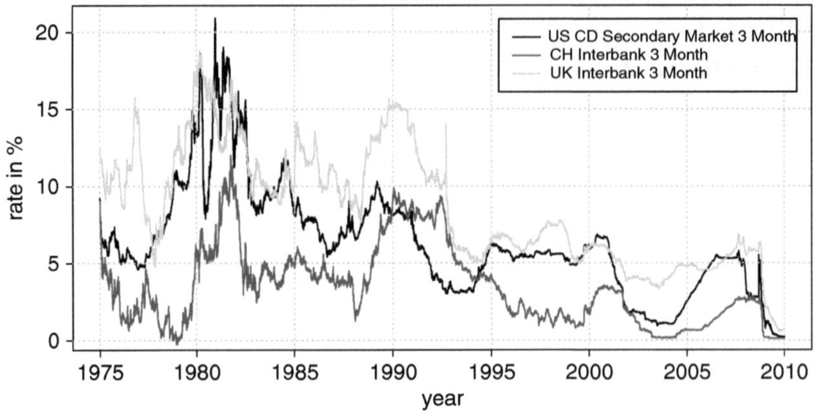

Fig. 1.1 A historic comparison of market interest rates form 1975–2009 (Source: Datastream)

In conjunction with embedded options and guarantees we often hear the history of *the Equitable Life Assurance Society* (Equitable Life). To report this I follow FSA (2001) and the content of Equitable Life's webpage.[1]

1.1 The History of Equitable Life

Founded in 1762 Equitable Life was known as the oldest mutual life assurance company in the UK. Due to its revolutionary actuarial practices in the early years and continuously progressive and innovative life assurance business, the company enjoyed an excellent reputation for many years.

A notable innovation of Equitable Life was the launching of the retirement annuity in 1957 that enabled a flexible pension for self-employed people. Later, in the 1980s and the early 1990s, their rapid growth continued. During that high market interest rate period they issued a big number of with-profit pension policies including a guaranteed annuity option (GAO). This option gives the policyholder the right to convert his accumulated funds to a life annuity at a fixed rate. The opposite of a GAO is the so called current annuity option (CAO) where the effective annuity rate is anticipated to current investment conditions. The value of the CAO falls if market interest rates fall or if the average life expectancy rises, and vice versa.

The GAO offered was at 4% before it was increased to 7% for the period between 1975 and 1988. At its peak around 175,000 of 1,100,000 Equitable Life's policyholders held a GAO in their policies. When market interest rates decreased in subsequent years and even the current rate fell below the guaranteed rate, the

[1] See http://www.equitable.co.uk/. Last checked: 1 June 2011.

GAO suddenly turned in the money, and as they did not recover in subsequent years the GAO severely threatened the financial stability of the company. Looking back we can say that obviously the then responsible management and its actuaries either were not able to appreciate the enormous financial risk that particular option bore or they just ignored it. Or with the unambiguous words of FSA (2001):

> It is apparent that at all material times from the 1980s onwards, Equitable Life was aware of the GAR risk (though it is not clear when the full possible impact of that risk was appreciated). It is also apparent that at no time did Equitable Life ever hedge or reinsure adequately against the GAR risk to counteract it.

As Equitable Life realised the dimension of their misinterpretation, they decided, according to one of its articles of association, to pay lower terminal bonuses to those policyholders who opted for exercising the GAO. The idea behind that decision is self-evident, a higher GAO rate applied to a lower total amount of benefits reduces the attractiveness of the GAO relative to the CAO on a higher total benefit. Hence less policyholders would exercise the GAO and prevent the company incurring extra costs. Of course, policyholders were not amused about that practice and took Equitable Life to court.

In July 2000 the British House of Lords decreed that Equitable Life could not set the level of terminal bonuses depending upon whether a GAO policyholder elected to take his benefits in GAO form or not.[2] Consequently Equitable Life had to establish additional money for outstanding claims of the past 7 years and even increased their financial difficulties.

The story came to an end when the board decided to put up Equitable Life for sale in late summer 2000. Since no buyer was found, they finally announced the ordinary run-off for their business on 8 December of the same year. After 238 years in business, one of the oldest and most renowned life insurer stumbled upon the GAO. As a mutual society, Equitable Life had some 425,000 members holding a £21 billion with-profits funds in the end.[3]

So what can we learn from this? There are three things I want to point out. First, we have seen how dangerous a single, inconspicuous type of an embedded option can be if insurers do not handle it with care. Especially the long term characteristics of this kind of option and guarantee is what makes it so hard to appraise its impacts.

Second, we have to realise that low market interest rates are problematic for two reasons. On the one hand the return on interest rate sensitive but usually less volatile assets such as e.g. bonds drops the lower the market interest rates are. On the other hand, I have outlined that many embedded options and guarantees are interest sensitive, too. But as we have seen with the GAO their value raises from the perspective of policyholders as market interest rates fall. This is also the case with many other embedded options and guarantees.

---

[2] See House of Lords (2000) for the full judgment.

[3] See House of Lords (2000).

The disaster of Equitable Life was caused, amongst other things, by human failure. But as a third point we also have to scrutinise the role of the then regulatory and accounting systems. Perhaps they have failed in that situation, too. However, by the beginning of the twenty-first century, when Equitable Life collapsed and many other insurance companies suffered serious financial distress a far-reaching change in the way of thinking of local insurance supervision took place all over the world. It was realised that current and mostly rule-based supervision could no longer guarantee an accurate protection of policyholder's interests. The change towards a more transparent, risk-based and dynamic supervision requesting for an active examination of one's own risk situation had begun.

Parallel to this process, new accounting and reporting standards were developed. Its impacts reached far beyond the insurance industry but affected the whole financial sector. Investors and analysts also urged the insurance industry to present new, appropriate and transparent approaches for measuring the value of their life business.

1.2 Revised Supervision and Valuation Methodology

These approaches do have a fundamental assumption in common with the revised supervision, the market consistent valuation of all assets and liabilities. In particular, this also relates to taking embedded options and guarantees explicitly into account.

Definition 1 (market consistent). The market consistent value of any security is equal to its market value (or market price) if there exists a deep and liquid market for the particular security. If there exists no such market then the market consistent value shall be equal to the present value of the security's best estimated cash flow discounted with appropriate market discounting factors.

Note that there is no generally accepted definition for market consistency and indeed it is not obvious here what is meant by a "deep and liquid market" or "appropriate market discounting factors". However, we will return to that issue in Chap. 3 by defining a market and discounting factors for modelling purposes.

To see what is the reason for requiring market consistent valuation, two of a number of advantages shall be pointed out.

- Since the valuation is based on observable market parameters, market consistent valuation can be deemed to be an objective measure.
- Market consistent valuation also brings consistency between the valuation of assets and liabilities.

In the following I will introduce the currently applied Swiss supervisory approach, called the *Swiss Solvency Test* (SST), and the European supervisory approach, called Solvency II, due to come into force 31 December 2012. Using the example of Equitable Life we have seen the meaning of embedded options and guarantees and so I will focus on its treatment when considering these approaches.

In addition to the new supervisory approaches, I will present a valuation framework that takes a shareholder's perspective, called the *Market Consistent Embedded Value* (MCEV). For my scope this is another interesting framework because embedded options and guarantees are explicitly taken into account, too. And as we will see there are many more similarities even if the main purpose of the MCEV is different than for the solvency frameworks.

1.2.1 The Swiss Solvency Test (SST)

The *Swiss Solvency Test* (SST) is the current solvency framework being applied in Switzerland. It was initiated in 2003 by the former Swiss insurance supervisory, the *Federal Office of Private Insurance* (FOPI), and continuously developed in cooperation with the insurance industry, consulting companies and academia. After a series of field tests the SST has been in force since 1 January 2006 and has become mandatory as of 2008 to all direct insurers, reinsurers and insurance groups doing business in Switzerland.[4] The legal basis for the SST can be found in VAG (2009) and AVO (2009). To introduce the main concepts of the SST on a general level, I will closely follow FOPI (2004a) in the subsequent. For a more technical view on the framework, the interested reader may have a look at FOPI (2004b).

Today the Swiss insurance supervisory is integrated in the *Swiss Financial Market Supervisory Authority* (FINMA). It is the goal of the FINMA then to ensure that the interests of any policyholders are protected. Former standard formulaic approaches used for calculating solvency requirements have turned out to be unsuitable for today's challenges. Nowadays, risk-based solvency standards to quantify the actual risks taken by the insurers seem to be an appropriate approach. So now it is the insurer's responsibility to actively investigate their own risk situation. Insurers shall be encouraged to do this properly by taking this into account in solvency capital calculations and probably lowering the individual minimum requirements. Note that this also enhances the transparency of the supervision and helps minimise systematic risks.

The aims of the renewed supervision are translated into a mathematical model, the SST. Find a brief overview on the structure of the SST in Fig. 1.2. It is a framework that is based on a series of standard models and a set of various scenarios. Due to the concept of risk-based supervisory, there is allowance for extending or replacing standard models by internal models. Additionally, each company is requested to complement predefined scenarios by company-specific ones.

Those models, standard or internal, shall capture all kind of insurance and financial risks. Therefore an important requirement is the market consistent valuation of assets and liabilities.

[4]The SST is not mandatory to foreign insurers only doing reinsurance business in Switzerland. Confer VAG (2009), 2.

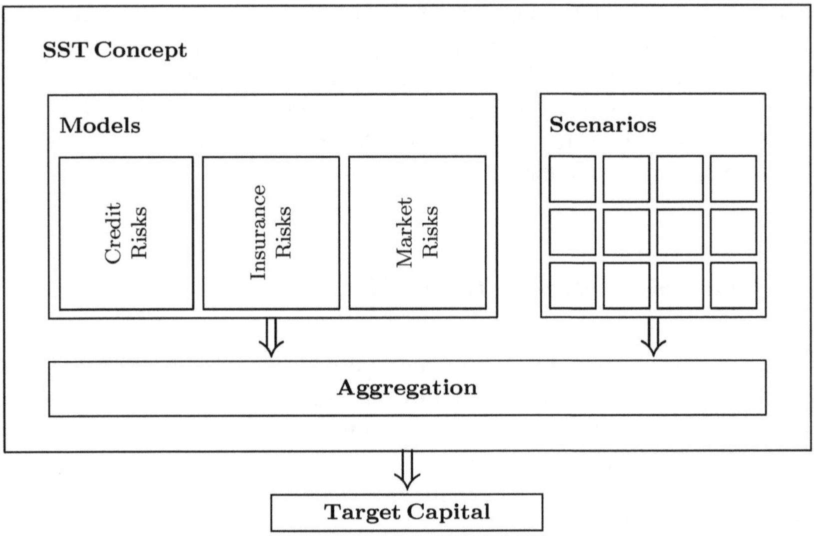

Fig. 1.2 The structure of the Swiss Solvency Test (Source: Rüfenacht/ FOPI (2004a))

One key figure to be calculated according to the SST modelling framework is the *target capital* (TC) that quantifies the minimum required capital for an insurance company. The TC ensures that even in a very unlikely situation there is enough capital on average to transfer the insurer's portfolio to a third party. Hence the TC has to suffice to cover future liabilities but also future costs of capital. From this definition it gets obvious that the SST also allows for defaults.

In addition to the required capital there is the available capital, called the *risk bearing capital* (RBC) that is defined as the difference between the market consistent value of the assets and the discounted best estimate cash flow of the liabilities. Obviously the RBC must be greater than the TC.

From today's RBC its 1 year ahead distribution is calculated. And from this distribution the TC will be derived.[5] So we see that the TC actually is a function of the RBC and that the general time horizon of the SST is 1 year.

Taking AVO (2009), the regulatory framework of the SST, we find one article that explicitly regulates taking relevant options and guarantees into account for calculating the TC, especially for life insurers:

AVO (2009) 46 I

Bei der Ermittlung des Zielkapitals, insbesondere in der Lebensversicherung, sind wesentliche Garantien und eingebettete Optionen zu berücksichtigen.

[5]For more detailed explanations see FOPI (2004b).

Some further explanations can be found in the appendix of AVO (2009). Considering the general code of practice for the common SST, FOPI (2004b), we do not find any explicit formulae to value the embedded options and guarantees but are referred to SAA (2006), published by the Swiss Association of Actuaries. Note that this is a guideline that only "aims to assist companies in valuing and modelling options and financial guarantees embedded in insurance products"[6] and there exists no code of practice.

1.2.2 Solvency II

Solvency II is a project to reform the current European supervision of insurance undertakings in the *European Union* (EU), known as Solvency I. Concluded in 2003 the *European Comission* (EC) published the *Solvency II Framework Directive Proposal* in July 2007 and an amended version containing conceptual information and rules in February 2008. Passed on 25 November 2010, the final directive EC 138 (2009) provides the regulatory framework for Solvency II. To see impacts and further improve the new solvency standard, the *Committee of European Insurance and Occupational Pensions Supervisors* (CEIOPS) initiated a series of *qualitative impact studies* (QIS). The latest study QIS 5 took place in late 2010. The interested reader can find detailed information on the different QIS on the *European insurance and reinsurance federation's* webpage.[7] Following current plans, Solvency II will be in force as of January 2013. In the following I will follow Swiss Re (2006) to introduce the main aspects of Solvency II.

The ultimate ambition of Solvency II is still to protect policyholder's interests. As a further objective a solvency measure reflecting an insurer's undertaken risks shall be provided. Together this leads to a new principle-based and transparent solvency standard with a focus on the entire solvency position of an insurance company. Indeed Solvency II is a quite ambitious project since it aims for a harmonisation of supervisory methods across the whole EU.

The entire concept of Solvency II is based on a three pillar concept. The first pillar handles quantitative requirements whereas qualitative requirements are handled in the second pillar. Finally market discipline regarding disclosure and transparency is treated in the third pillar. A more detailed explanation of each pillar's content follows in the subsequent.

---

[6] SAA (2006), p. 2.

[7] http://www.cea.eu/index.php/key-issues/solvency-ii/ceiops-quantitative-impact-studies. Last checked: 1 June 2011.

Pillar 1 – The Head

In Pillar 1 the quantitative requirements of Solvency II are covered. It defines the financial resources that a company needs to keep solvent. Therefore two key quantities are to be calculated on a market consistent valuation basis, the *solvency capital requirement* (SCR) and the *minimum capital requirement* (MCR), that mark some sort of thresholds.[8]

The SCR can be calculated either according to a standard formula or based on internal models whereas Solvency II clearly encourages insurance companies to develop and to use internal models.[9] Given the SCR an insurer should be able to absorb significant and unforeseen losses and still assure their policyholder's claims. More technically spoken, the SCR quantifies the required capital to meet all obligations over a given time horizon to a given confidence level. Thus, within these calculations, all significant risks and quantifiable risks an insurer is exposed to should be taken into account. Once a company's available capital falls below the SCR, supervisory action according to Pillar 2 will be triggered.

In comparison, the MCR is the lower threshold. If the available capital sinks below that level, the company presents an unacceptable risk to its policyholders. In such a situation the supervisory authority can invoke severe measures according to Pillar 2 again and even close a company to new business. Guidelines for calculating the MCR are given in articles 128 and 129 of the directive EC 138 (2009).

Pillar 2 – The Heart

In Pillar 2 the qualitative requirements of Solvency II are defined. It provides principles for the supervisory control and for an insurer's internal risk management. As we have seen in the previous paragraph, certain rules in case of a precarious financial situation identified under Pillar 1 are outlined in this pillar here. Furthermore Pillar 2 should asses risks that cannot be quantified within Pillar 1.

Pillar 3 – The Mouth

The last pillar covers issues on disclosure and supervisory reporting. It is the goal of Pillar 3 to give all parties involved such as policyholders, investors and further a comprehensive picture of an insurer's risk exposure. It is also intended to assist transparency and to bring market discipline.

In Fig. 1.3 the key elements of the three pillar concept are summarised.

---

[8] Note that market consistency is also a standard assumption within Solvency II but with the same lack of definition as for the SST. Taking the directive EC 138 (2009) we find: "(54) The calculation of technical provisions should be consistent with the valuation of assets and other liabilities, market consistent and in line with international developments in accounting and supervision."

[9] See Annex IV of the directive EC 138 (2009) for the *SOLVENCY CAPITAL REQUIREMENT (SCR) STANDARD FORMULA*.

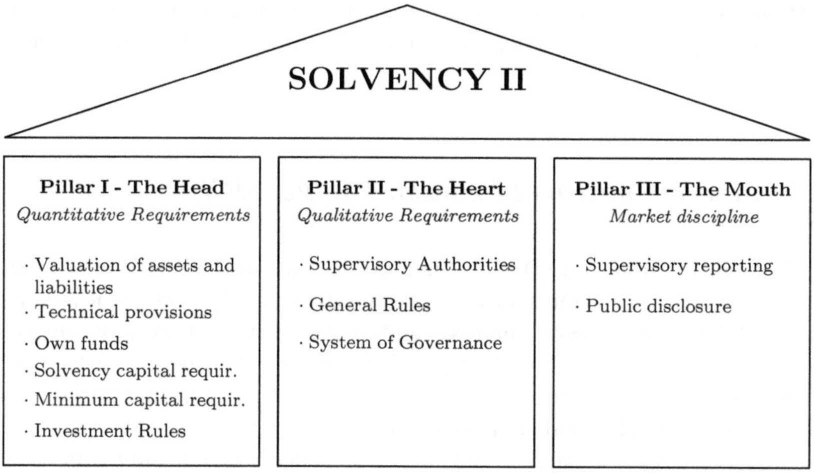

Fig. 1.3 The three pillars of Solvency II

Embedded options and guarantees are also an issue within Solvency II. In order to begin we find that they are one of the points of motivation for the revised directive in EC 138 (2009) itself.

(58), EC 138 (2009)

> It is necessary that the expected present value of insurance liabilities is calculated on the basis of current and credible information and realistic assumptions, taking account of financial guarantees and options in insurance or reinsurance contracts, to deliver an economic valuation of insurance or reinsurance obligations. The use of effective and harmonised actuarial methodologies should be required.

Hence, they have recognised the importance of taking embedded options and guarantees into account. To continue there is one article explicitly covering the "valuation of financial guarantees and contractual options included in insurance and reinsurance contracts":

EC 138 (2009), 79

> When calculating technical provisions, insurance and reinsurance undertakings shall take account of the value of financial guarantees and any contractual options included in insurance and reinsurance policies.
> Any assumptions made by insurance and reinsurance undertakings with respect to the likelihood that policy holders will exercise contractual options, including lapses and surrenders, shall be realistic and based on current and credible information. The assumptions shall take account, either explicitly or implicitly, of the impact that future changes in financial and non-financial conditions may have on the exercise of those options.

Guidelines on how to apply that article 79 are outlined in the technical specifications for the fifth quantitative impact study QIS 5 CEIOPS (2010a) and its annexes CEIOPS (2010b).

1.2.3 Market Consistent Embedded Value (MCEV)

The *Embedded Value* (EV) is a measure to quantify the value of the life insurance business companies have with their customers. This concept was mainly introduced for shareholders' interests and is not requested by any supervisory authority. However EV methodology provides several interesting aspects regarding to embedded options and guarantees and thus allowance for a brief introduction to the EV concept shall be made in the current section.

In recent years the insurance industry has recognised that only a transparent, consistent and comparable EV can become a meaningful valuation standard. That is the goal the *European Chief Financial Officer Forum* (CFO Forum) wanted to achieve when publishing their *Market Consistent Embedded Value Principles*© (MCEV Principles) in June 2008.[10] Its latest version is written down in CFO Forum (2009a). The interested reader may also have a look at CFO Forum (2009b) that provides supporting rationale for the MCEV Principles.

The MCEV is defined as follows in the MCEV Principles[11]:

$$\begin{aligned}\text{MCEV} = \ &\text{Free Surplus Allocated to the Covered Business}\\ &+\text{Required Capital}\\ &+\text{Value of In-Force Covered Business (VIF)}\end{aligned}$$

whereas

$$\begin{aligned}\text{VIF} = \ &\text{Present Value of Future Profits}\\ &-\text{Time Value of Financial Options and Guarantees}\\ &-\text{Cost of Residual Non-Hedgeable Risks}\\ &-\text{Fricitional Cost of Required Capital}\end{aligned}$$

Hence the MCEV is basically represented by the summed value of a couple of single entities. The precise meaning of each of these entities is outlined in the MCEV principles. There we find that the free surplus is the market value of any assets allocated to, but not required to support, the in-force covered business

[10] See http://www.cfoforum.eu/. Last checked: 1 June 2011.

[11] See Principles 3–9 in CFO Forum (2009a).

at the valuation date. Besides the required capital is the market value of assets, attributed to the covered business over and above that required capital to back liabilities for covered business, whose distribution to shareholders is restricted. Note that the influence of market consistency within the MCEV methodology is clearly observable in these two definitions already.

Very often we find the notion of the *adjusted net asset value* (ANAV) that puts these two positions together. From the remaining positions the focus shall be on the embedded options and guarantees which are covered in Principle 7.

CFO Forum (2009a), Principle 7

> Allowance must be made in the MCEV for the potential impact on future shareholder cash flows of all financial options and guarantees within the in-force covered business. The allowance for the time value of financial options and guarantees must be based on stochastic techniques using methods and assumptions consistent with the underlying embedded value. All projected cash flows should be valued using economic assumptions such that they are valued in line with the price of similar cash flows that are traded in the capital markets.

By requesting the application of stochastic techniques this principle appears to show quite concretely how to handle embedded options and guarantees. In the sense of a principles-based methodology it is again the insurer's problem to develop an appropriate framework. Further guidance is given in CFO Forum (2009b).

1.2.4 Comparison of the Introduced Approaches

To close this chapter I would like to give the reader a brief summary of the concepts introduced until now by comparing some selected topics.

SST	Solvency II	MCEV
Aims		
Protection of policyholders' interests. Protection of the financial system's stability only indirectly recognised	Protection of policyholders' interests and the stability of the financial system as a whole. Furthermore it aims for a harmonisation of solvency requirements across the EU	Providing a transparent and consistent measure of the value of shareholders' interests in the covered business

(continued)

SST	Solvency II	MCEV
Coverage		
Mandatory to all insurers doing life, non-life or reinsurance business in Switzerland as of 2008, solvency according to SST to be achieved as of 2011	Will become mandatory to all insurers doing life, non-life or reinsurance business in the EU following current plans as of 2013	MCEV Principles were published in 2008 but it is not mandatory for any insurer. However MCEV is requested by investors and analysts for larger life insurance companies
Risks considered		
All kind of significant and quantifiable insurance and financial risks an insurer is exposed to excluding operational risks	All kind of significant and quantifiable insurance and financial risks an insurer is exposed to including operational risks	Hedgeable financial risks, the frictional costs of required capital and the cost of residual non-hedgeable risks
Valuation principles		
Market consistent valuation of all assets and liabilities	Market consistent valuation of all assets and liabilities	Market consistent valuation of all assets and liabilities
Quantitative requirements		
· Target capital · Risk bearing capital	· Solv. capital requirement · Min. capital requirement	· Adjusted net asset value · PV of future profits · Time val. of options and guarantees · Cost of residual non-hedgeable risks · Frictional costs of capital
Reference rate		
Gov. bond yield curve	Swap yield curve	Swap yield curve
Main risk measure		
Expected shortfall	Value at risk	Expected value

(continued)

SST	Solvency II	MCEV
Qualitative requirements		
Given standard models with allowance for internal models. Fully principles-based and the development and use of internal models is desired	Given standard models with partly allowance for internal models. Mostly principles-based	MCEV methodology is fully principles-based but there exists no standard model
Experiences		
Field tests during 2004–2005	Quantitative impact studies (QIS 1–QIS 5) since 2005	Embedded value and European Embedded Value (EEV) disclosures from earlier years
Embedded options and guarantees		
Embedded options and guarantees are to be taken into account for valuing the insurance liabilities. Guidelines provided in a paper by the SAA	Embedded options and guarantees are to be taken into account for valuing the insurance liabilities. Guidelines provided in QIS 5	Embedded options and guarantees are to be taken into account for valuing the insurance liabilities
Literature		
· FOPI (2004a) · FOPI (2004b)	· EC 138 (2009) · CEIOPS (2010a) · CEIOPS (2010b)	· CFO Forum (2009a) · CFO Forum (2009b)

1.3 Research Question and Objectives

We have seen how threatening embedded options and guarantees can become for the financial stability of an insurance company. Most of these embedded options and guarantees are sensitive to market interest rates, whereas its value increases from a policyholder's perspective as market interest rates decrease. Indeed this is what could have been observed during the last two to three decades on the major markets in Europe and the United States.

Due to critical years for the insurance industry caused by failing financial markets, increased longevity and low market interest rates, a far-reaching change in the way of thinking of local insurance supervisory and general accounting methodology took place all over the world at the beginning of the twenty-first century. The SST for Switzerland and Solvency II for the EU are the two examples I have presented in this context. Additionally I have given an introduction to the MCEV methodology that is rather a valuation framework than a solvency framework but nevertheless has many issues in common with the SST and Solvency II, especially regarding to handling embedded options and guarantees. So it shall be this set of theoretical frameworks which provide the qualitative background of my thesis' motivation.

Indeed, knowing the value of embedded options and guarantees is of vital importance for the financial stability of a life insurance company. Since there exists no market and hence no observable market value for this, I identify the need for a quantitative modelling approach to handle these embedded options and guarantees based on modern valuation techniques with respect to current regulatory and accounting standards. It is important to recognise that here market consistency is the basic prerequisite for current applied valuation methodologies and so this model must be set up. For this reason it shall be calibrated consistently to true market data and allow for considering real, existing and in-force policy data. At the same time I will clearly focus on the modelling of interest rates.

The central question I aim for answering with this model is the following: What value do the considered embedded options and guarantees have, absolutely and relative to the current policy reserves, given a perspective taken from the insurer, the policyholder and the shareholder respectively? With respect to the modelling, taking someone's perspective is equivalent to optimising certain variables under given constraints. On the one hand, the insurer intends to minimise losses in due consideration of competitive dividend and surplus payments. On the other hand, the policyholder is only interested in its guaranteed benefits and maximised surpluses and the shareholder of course wants to have maximised and stable dividend payments.

By enabling to take an insurer's, a policyholder's or a shareholder's perspective this model finally provides a great flexibility and hence forms a useful analysis tool. Over and above the values of the considered embedded options and guarantees, the results shall support actuaries as well as risk managers defining an optimal dividend and surplus policy and improving their asset liability management. These are consequently the ultimate goals of my thesis.

1.4 Outline

The remainder of this thesis is organised as follows. Chapter 2 specifies the terms *embedded option* and *guarantee* and presents a systematisation of common examples thereof. Subsequently there will be a literature survey to outline the reader the actual state of the scientific knowledge that is of relevance for the purpose of the thesis. In Chap. 3 the theoretical background of the asset modelling process is defined. This asset modelling process is supposed to consist of three stochastic processes: A share price process, a property price process and a zero coupon bond price process. After a proper introduction to stochastic modelling, each of these price processes shall be described and in the following calibrated to observable market data. Chapter 4 outlines the available insurance specific data for modelling purposes. Furthermore the entire modelling algorithm is completed by presenting the liability modelling process. The climax of this thesis is presented in Chap. 5: An extensive empirical analysis based on a large set of existing in-force insurance policies using the recently developed entire modelling algorithm. The focus is on three different options, the interest rate guarantee, the dividend option and the shareholder option. To determine their market consistent values, the modelling algorithm is performed using consequently best estimate assumptions first. This is called the basic simulation run. Then, taking the results of the basic simulation run as reference values, a detailed sensitivity analysis is carried out. Finally, Chap. 6 concludes and provides an outlook for further research opportunities on this topic.

Chapter 2
Theoretical Considerations Regarding Embedded Options

In the previous chapter I have introduced the subject of embedded options and guarantees. Now I will specify the terms *embedded option* and *guarantee* and present a systematisation of common examples thereof. Subsequently there will be a literature survey to outline to the reader the actual state of scientific knowledge that is of relevance for the purpose of this thesis.

Before starting, some delimitation are in order. Any kind of embedded option or guarantee is always linked to a particular life insurance contract. Within this thesis, neither non-life nor reinsurance business will be considered. Furthermore, only financial options and financial guarantees are taken into account.

Roughly speaking a *life insurance* is merely an umbrella term for any kind of insurance contract where the benefit is linked to the policyholder's life and risks as survival and death are covered. Note that within this thesis a policyholder of a life insurance is always the insured person as well. In the main there are three life insurance base products.

Term Life Insurance

A term life insurance covers the insured person for a fixed period of time (term) that is shorter than lifetime. If the insured person dies during the term a death benefit is due.

Whole Life Insurance

A whole life insurance covers the insured person for life. A death benefit is due when the insured person dies.

N. Rüfenacht, *Implicit Embedded Options in Life Insurance Contracts*, Contributions to Management Science, DOI 10.1007/978-3-7908-2843-6_2,
© Springer-Verlag Berlin Heidelberg 2012

Endowment Insurance

An endowment insurance covers the insured person for a fixed period of time (term) that is shorter than lifetime. If the insured person dies during the term a death benefit is due. If the insured person survives the term a survival benefit is due.

For a deeper insight into life insurance methodology and basic life insurance mathematics I recommend Gerber (1997).

There are many possibilities to vary these base products. Different premium payment schemes, increasing death or survival benefits and extended coverage are just a few examples to specify a particular life insurance contract. As we have already seen an insurer might further add certain embedded options and guarantees to its insurance policies. But what are these embedded options and guarantees precisely? In the classical financial market theory an option is defined as follows:

Definition 2 (Call/Put Option). A call/put option is a contract that gives the holder the right, but not the obligation, to buy/sell a specified security (the underlying) by a certain date (the exercise date) for a certain price (the strike price). An American option can be exercised at any time up to the expiration date whereas European options can be exercised only at the expiration date itself.

For a general introduction to option theory and especially option pricing theory I refer the interested reader to Hull (2006). However this is a natural point to define what an embedded option shall be for the purpose of this thesis. With some modifications we get the following:

Definition 3 (Embedded Option in a Life Insurance Contract). An option linked to a life insurance contract is called an embedded option (in a life insurance contract). It gives the policyholder the right, but not the obligation, to actively change his or her life insurance contract by a certain date on terms that are established in advance. European and American type of options are feasible.

2.1 Types of Embedded Options and Guarantees

By way of illustration it is most intuitive to give some tangible examples of embedded options. This selection is non-exhaustive of course but covers the most material embedded options that appear quite frequently in common life insurance products. For this I will follow CEIOPS (2010a).

Surrender Option

A surrender option gives the policyholder the right to fully or partially surrender the policy and receive a pre-defined lump sum amount, the surrender value. That value usually consists of the policy reserve including probable dividends but reduced by a surrender fee. Normally the surrender option can be exercised by the end of any contract year. Thus, in theory, it is a Bermudan put option on future expected insurance benefits with a strike price equal to the surrender value.[1] In Switzerland for example the surrender option is requested by law for life insurance contracts with a savings character.[2]

Paid-up Policy Option

A paid-up policy option gives the policyholder the right to stop paying premiums. The current policy continues with reduced benefits depending on the policy reserves present at the exercise date. This right can be seen as a put option with a strike price equal to zero on a forward rate agreement. De facto the paid-up policy option exchanges the original contract for a new contract with reduced benefits and no more premium payments. Similar to the surrender option the paid-up option is requested by law in Switzerland, nevertheless the latter option is required for all types of life insurance contracts.[3]

Annuity Conversion Option

An annuity conversion option gives the policyholder the right to convert a lump survival benefit into an annuity at a pre-defined minimum rate of conversion. This option is also known as the GAO that has been introduced in the previous chapter. By converting the policyholder exercises an European call option at a strike price equal to the survival benefit on a lifelong annuity.

Policy Conversion Option

A policy conversion option gives the policyholder the right to convert from the current policy to another at pre-specified terms and conditions. For example a term insurance policy could be converted into an endowment policy. Typical of this option is that it may be exercised without further evidence of insurability.

So far we have seen that most of these embedded options can be described in terms of the classical option theory. Nevertheless, this is rather artificial. However for the purpose of this thesis the crucial aspect is the behaviour of the policyholder since it is his decision whether to exercise any particular option or not.

There exists many more types of embedded options, being in my opinion less relevant for the insurance industry than the ones presented above. However for a

---

[1] A Bermudan (call or put) option has predetermined, discrete exercise dates. Hence it is something between American and European option type.

[2] VVG (2009) 90 II.

[3] VVG (2009) 90 I.

more extended overview on embedded options I refer the interested reader to Gatzert (2009).

Besides embedded options, life insurance policies are offered containing certain guarantees enabling the policyholder to avert losses or to receive additional benefits. A guarantee is usually applied automatically as defined in the policy terms and conditions and always rebounds to a policyholder's advantage. To remain consistent with the former definition of an embedded option we simply consider guarantees as *implicit* embedded options without the feature of the policyholder's right to actively exercise the option.

To present the most relevant guarantees offered I will follow CEIOPS (2010a) again.

Minimum Interest Rate Guarantee

With a minimum interest rate guarantee savings premiums are accumulated with that pre-defined minimum interest rate. The minimum interest rate is also termed as the technical interest rate. In Switzerland the federal council dictates the maximum height of the technical interest rate for new business.[4] Currently the technical interest rate is bounded above at 1.75% for policies in Swiss francs, at 2.5% for policies in Euro and at 2.75% for policies in US dollars.[5]

Dividend Option

A dividend option enables policyholders to partake of the profits gained by the insurance company due to positive investment, mortality and expenses results. There are several possibilities of implementing the dividend option in practice. The insurer can either pay out dividends as cash, reduce future premiums or accumulate with interest. Dividends allocated to policyholders are also called surpluses. A life insurance policy containing a dividend option is also called a *participating life insurance policy*.

Guaranteed Minimum Accumulation Benefit

Unit-linked life insurance products "link" the policyholder's benefit to some financial index or fund. Savings premiums are continuously invested in the underlying and so the benefits vary according to its performance. To avoid too low benefits relative to the savings premiums paid due to a weak or even negative performance of the underlying, the insurer may guarantee a minimum accumulation benefit. Similarly a guarantee for a minimum death benefit is possible.

---

[4]Confer VAG (2009) 36 I and AVO (2009) 121.

[5]See http://www.finma.ch/e/beaufsichtigte/versicherungen/lebensversicherungen/Pages/max-tech-zinssatz.aspx. Last checked: 1 June 2011.

There are a few more examples of guarantees within life insurance contracts in Gatzert (2009) again. After having presented the most material embedded options and guarantees, I will proceed by defining those options the focus will be on within the empirical analysis in Chap. 5. Based on that the literature survey will be presented.

Basically I aim to design a fairly general framework that provides an appropriate substructure to handle every kind of embedded option and guarantee. Since there is a focus put on selected options and guarantees the framework finally being presented and applied for modelling purposes in the subsequent chapters will be some special case thereof. But note that this does not affect the requested flexibility of the model itself.

A model is always a simplification of the real world. Nevertheless one should plan for mapping at least the key elements of any model as realistically as possible. With regard to embedded options probably the main key factor is the behaviour of policyholders. This is also recognised by the new solvency and valuation standards and so we find for example the following for the SST.

SAA (2006), Sect. 2.2

> (...) It is vital to be consistent when including policyholder and insurer options in solvency calculations. This requires realistic modelling of both policyholder and insurer behaviour. (...)

SAA (2006) does not further explain what *realistic modelling of policyholder behaviour* means exactly. For the European Solvency II regime we find quite a similar instruction in Article 79 of the directive EC 138 (2009) that I have cited in Sect. 1.2.2 already. Assumptions made with respect to the likelihood that policyholders will exercise embedded options *shall take account, either explicitly or implicitly, of the impact that future changes in financial and non-financial conditions may have on the exercise of those options.*

In addition CEIOPS (2010a) addresses explicitly to the policyholders' behaviour in TP.2.105–TP.2.111. There are two paragraphs I consider worth noting here.

CEIOPS (2010a), TP2.107

> Assumptions about the likelihood that policy holders will exercise contractual options should be based on analysis of past policyholder behaviour. The analysis should take into account the following:
>
> (a) How beneficial the exercise of the options was or would have been to the policyholders under past circumstances (whether the option is out of or barely in the money or is in the money),
> (b) The influence of past economic conditions,

(c) The impact of past management actions,
(d) Where relevant, how past projections compared to the actual outcome,
(e) Any other circumstances that are likely to influence a decision whether to exercise the option.

CEIOPS (2010a), TP2.109

> In general policyholders' behaviour should not be assumed to be independent of financial markets, of undertaking's treatment of customers or publicly available information unless proper evidence to support the assumption can be observed.

These paragraphs outline the high requirements for an appropriate modelling of policyholder's behaviour. To do so, at least a profound statistical analysis of an insurer's own data would be necessary. Even though the behaviour of policyholders is of vital importance for modelling embedded options it will not be further quantitatively considered within this thesis. That issue is a self-contained one but lies beyond the scope of the thesis. So only guarantees in traditional life insurance products will finally be modelled, interest rate guarantee and dividend option.[6] Note that neither the goals of the thesis nor the flexibility of the model are diminished by omitting policyholder behaviour.

Finally we shall have a look at the MCEV framework. Some respective explanations are written down in the MCEV Principles:

CFO Forum (2009a), G7.2

> Where management discretion exists, has passed through an appropriate approval process and would be applied in ways that impact the time value of financial options and guarantees, the impact of such management discretion may be anticipated in the allowance for financial options and guarantees but should allow for market and policyholders' reaction to such action. (...)
> The application of such discretion must consider the environment arising in the future projection which will likely be different from the current environment, but any changes from current decision rules (for example regarding flexible crediting rates or policyholder bonuses) must be supported by appropriate approvals.

CFO Forum (2009a), G7.3

> Dynamic policyholder behaviour should, where material, be in the allowance for the time value of financial options and guarantees.

[6] With *traditional* life insurance products I mean non unit-linked products.

2.2 Literature Survey

The literature survey shall outline for the reader the actual state of the scientific knowledge. Note that all the papers to be discussed will be presented in the nomenclature of the respective authors.

As stated in the previous section, the focus within this thesis will be on the minimum interest rate guarantee and the dividend option, that is de facto a guarantee, too. Both guarantees have been applied for considerable length of time and nowadays these are taken for granted. One of the first papers addressing this issue was written by Briys and de Varenne.[7]

Briys and de Varenne (1997)

The motivation of the paper is based on a brief comparison of banking and insurance industry. It points out the long maturity of insurance liabilities as the main distinguishing feature implying remarkable challenges regarding the insurance industry's asset liability management (ALM). From insurance agents there is a considerable pressure to offer policies at a high guaranteed interest rate for competitive reasons. Keeping the rate high even when market interest rates fall, urges insurance companies to turn away from perfectly matching their long-term liabilities and therefore invest in shorter-term assets.

The model presented is based on Merton's approach to financial intermediaries and only one kind of artificial life insurance policy including interest rate guarantee and dividend option is considered.[8] This policy is comparable to a pure endowment insurance policy, an endowment insurance policy that pays a survival benefit only and no death benefits, paid for a single premium. Within the authors' words *the model enables one to determine the fair interest rate or the fair participation level policyholders should require to fully compensate them for the risks they face.*

Let $[0, T]$ be the planning horizon of any life insurance company. $t = 0$ shall be the time of issuance for a given cohort of life insurance policies maturing in $t = T$. At time $t = 0$ policyholders pay a single premium. Together with paid-in capital E_0 an asset portfolio A_0 is financed.

The life insurance policy shall guarantee a fixed yearly interest rate r_G to the policyholder. On top of this the insurance company shares a δ of its net financial revenues to its policyholders. Since r_G is usually lower than the market rate for a risk free asset of the same maturity as the policy, the participation coefficient δ is an important feature expressing the required risk premium for holding risky life insurance policies.

---

[7] See Briys and de Varenne (1997).

[8] Confer Merton (1977) and Merton (1990).

At start the company's balance sheet looks as follows (Table 2.1).

Table 2.1 The company's balance sheet at $t = 0$ in the Briys and de Varenne model (Source: Briys and de Varenne (1997))

Assets		Liabilities and equity	
Assets	A_0	Liabilities	$L_0 = \alpha \cdot A_0$
		Equity	$E_0 = (1-\alpha) \cdot A_0$
Total	A_0	Total	A_0

The term $(1-\alpha)$ gives the proportion of the initial assets financed by equity. The asset portfolio is assumed to consist of risky assets like equity, bonds or property only. The time t value of the asset portfolio A_t is then given by a stochastic process capturing both interest rate risk and asset risk.[9]

Over time stakeholder's cash flows occur as contractually defined. Then in $t = T$ three different states are feasible.

1. Total Insolvency

   The time T value of assets A_T is less than the guaranteed benefits and the company is declared bankrupt. The terminal benefit to policyholders is equal to the remaining amount of assets.

$$L_T = A_T \tag{2.1}$$

2. Partial Insolvency

   The company's assets are sufficiently high to pay guaranteed benefits but not to pay any bonus. Let B_T be the financial bonus to policyholders after guaranteed benefits were cashed out. Then

$$\begin{aligned} B_T &= \delta \left[\frac{L_0}{A_0}(A_T - A_0) - (L_T^g - L_0) \right] \\ &= \delta[\alpha A_T - L_T^g], \end{aligned} \tag{2.2}$$

   where δ denotes the level of bonus and $L_T^g = \alpha A_0 \exp(r^g \cdot T)$ is the guaranteed benefit to the policyholders. Obviously in this state the company's assets A_T are greater than the guaranteed benefits L_T^g. But if $A_T < L_T^g/\alpha$ there is no bonus since $B_T < 0$.

3. No Insolvency

   The company is able to pay guaranteed benefits and bonus to its policyholders. The total Liabilities at time T are given by

[9] The stochastic process is defined by the differential equation $dA_t/A_t = \mu dt + \sigma_A \cdot [\rho dW_t + \sqrt{1-\rho^2} dZ_t]$ where μ and σ_A denote the instantaneous expected return on assets and their instantaneous volatility respectively. W_t and Z_t are standard Wiener processes.

$$L_T = L_T^g + B_T \tag{2.3}$$
$$= (1-\delta)L_T^g + \delta\alpha A_T$$

Equity has a limited liability feature and thus shareholder's stake is a residual stake. Analogous to the total benefits the total equity payoff is depicted for each of the three different states.

1. Total Insolvency

$$A_T < L_T^g \Rightarrow E_T = 0 \tag{2.4}$$

2. Partial Insolvency

$$L_T^g \leq A_T < \frac{L_T^g}{\alpha} \Rightarrow E_T = A_T - L_T^g \tag{2.5}$$

3. No Insolvency

$$A_T \geq \frac{L_T^g}{\alpha} \Rightarrow E_T = (1-\delta\alpha)A_T - (1-\delta)L_T^g \tag{2.6}$$

Liability and equity payoffs depending on the different states are summarised in Fig. 2.1.

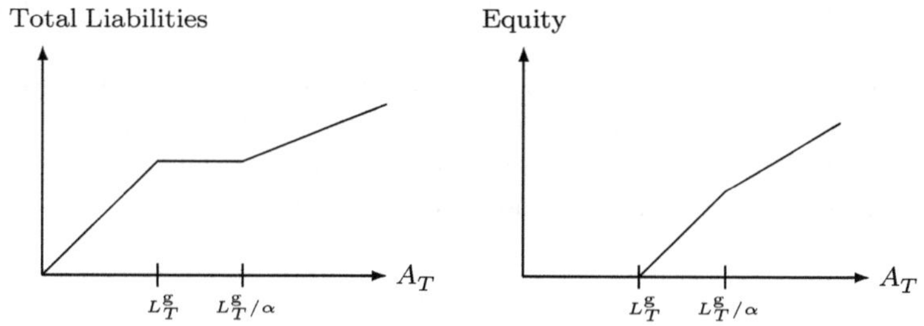

Fig. 2.1 Policyholder's and shareholder's final payoffs in the Briys and de Varenne model (Source: Rüfenacht/Briys and de Varenne (1997))

This framework is the starting point Briys and de Varenne founded their analysis on. They use a contingent claim based methodology to derive a closed-form solution for the pricing of life insurance liabilities. The authors also prove that the model produces more accurate duration and convexity measures since it accounts for stochastic interest rates, default risk and bonus schemes. A significant difference between the effective insurance liability and the traditional Macaulay duration is shown.

There are several suggestions to improve the model. Extensions to more complex life insurance products or taking surrender into account are conceivable for example.

Grosen and Jørgensen (2000)

In their introduction Grosen and Jørgensen expose the importance of a proper valuation and separate balance sheet reporting of embedded options. They mention interest rate guarantee, surrender option and dividend option as common examples embedded in many life insurance policies. Historically insufficient attention has been paid to accurate valuation or reporting issues. Several possible explanations are given for this, such as companies that were not realising or simply not caring about potential financial impacts caused by embedded options.

Motivated by insurance companies having suffered financial distress due to underestimating their embedded options, the authors present a multi-period model providing a smoothing surplus distribution mechanism. Basically that model is an extension of the one presented in Briys and de Varenne (1997). The main contribution of their work lies at a credible modelling of asset returns, a realistic bonus distribution scheme integrating the interest rate guarantee and lastly arbitrage-free valuation of the path-dependent contractual payoffs arising from the applied bonus distribution scheme.

The modelling framework is embedded in a continuous and frictionless economy with perfect financial markets. Thus transaction costs, liquidity squeezes, tax effects and other market imperfections will be ignored. Furthermore any kind of expenses, lapses and mortality are ignored, too.

Let $[0, T]$ be the considered time horizon of the model. In $t = 0$ the policyholder pays a single-sum deposit V_0 to the insurance company and therefor acquires a policy of nominal value P_0. In $t = T$ the policy matures and the policyholder receives a single payment.

When the policy is issued in $t = 0$, V_0 is invested in a portfolio with traded assets at the financial markets, say the *reference portfolio*. Its initial value is denoted by A_0. To illustrate the modelling algorithm a simplified time t balance sheet is used as shown in Table 2.2. Note that this is not a company's balance sheet.

Table 2.2 The policy's time t balance sheet in the Grosen and Jørgensen model (Source: Grosen and Jørgensen (2000))

Assets	Liabilities
$A(t)$	$P(t)$
	$B(t)$
$A(t)$	$A(t)$

$A(t)$ denotes the time t market value of the reference portfolio backing the policy liability. On the liability side we see $P(t)$, the *policyholder's account* or *policy reserve*. $B(t)$ stands for the *bonus reserve* or simply the *buffer*. On the one hand A is clearly a market value whereas P is rather a book value on the other hand. Finally B is defined as a residual quantity given by the difference $A - P$ and thus a hybrid of a market value and a book value. The authors emphasise that equity is not a priori missing in their model but implicitly included as a part of the bonus reserve.

The reference portfolio is assumed to evolve according to the following geometric Brownian motion.

$$dA(t) = \mu A(t)dt + \sigma A(t)dW(t), \qquad A(0) = A_0, \tag{2.7}$$

where μ and σ denote the instantaneous expected return on assets and their instantaneous volatility respectively. $W(t)$ is a standard Wiener process.

The dynamics of the policy reserve P is determined by the sequence $r_P(t)$, the *policy interest rate*, where $t \in I = \{1, 2, \ldots, T\}$. In particular $P(t)$ is defined recursively by

$$P(t) = (1 + r_P(t)) \cdot P(t-1), \qquad t \in I, \tag{2.8}$$

that leads to

$$P(t) = P_0 \prod_{i=1}^{t} (1 + r_P(i)), \qquad t \in I. \tag{2.9}$$

In practice determining r_P is highly subtle influenced by political, legal or strategic considerations. A priori one can decompose the policy interest rate into the sum of the *guaranteed interest rate* r_G and some *bonus interest rate* r_B.

$$r_P(t) = r_G + r_B(t), \qquad t \in I. \tag{2.10}$$

For this model's purposes a simplification of the precise and highly sophisticated process in practice to determine the bonus interest rate r_B and accordingly the policy interest rate r_P is assumed. Here some features are requested. Firstly, realised returns on assets must influence the future policy interest rate r_P. Secondly, r_P should be lowly volatile and yet competitive compared to other market assets of equal riskiness. Thirdly, the insurance company aims for maintaining a certain buffer in order to provide stable returns to its policyholders. So summarising this arguments yields in a policy interest rate r_P that depends on the guaranteed interest rate r_G, the performance of the reference portfolio A and the level of the buffer B.

Translated into mathematical terms the authors present the following term for the policy interest rate:

$$r_P(t) = \max\left\{ r_G, \alpha \cdot \left(\frac{B(t-1)}{P(t-1)} - \beta \right) \right\}, \tag{2.11}$$

where α is the *distribution ratio* and β the *target buffer ratio.* Using Formula 2.10 and 2.11 the bonus interest rate r_B can be expressed as follows.

$$r_B(t) = \max\left\{ 0, \alpha \cdot \left(\frac{B(t-1)}{P(t-1)} - \beta \right) - r_G \right\} \tag{2.12}$$

Recalling Formula 2.8 the evolution of the policyholder's account can be specified.

$$P(t) = P(t-1)\left(1+\max\left\{r_G, \alpha \cdot \left(\frac{B(t-1)}{P(t-1)} - \beta\right)\right\}\right) \tag{2.13}$$

$$= P(t-1)\left(1+r_G+\max\left\{0, \alpha \cdot \left(\frac{A(t-1)-P(t-1)}{P(t-1)} - \beta\right) - r_G\right\}\right)$$

At this stage there are two points worth noting. Firstly, the interest rate guarantee implies a floor on the terminal payout given by $P_{\min}(T) = (1+r_G)^T \cdot P_0$. It becomes effective if there is never any bonus distributed to the policyholders. Besides, the floor is lifted as soon as there is any bonus distributed since bonus is allocated to the policy account. Secondly, Eq. 2.13 implies that the bonus mechanism is indeed an embedded option.

To give the reader an impression of the model's behaviour two sample plots are in order. Figure 2.2 clearly demonstrates the smoothing effect on the policy interest rate r_P. The return on the reference portfolio A is quite volatile compared to r_P that is more or less stable.

In Fig. 2.3 the evolution of all relevant accounts modelled are plotted. Again it is a sample simulation with arbitrary set parameters but anyhow contains a lot of information on the model. Due to the interest rate guarantee the policy account P is strictly increasing whereas for the reference portfolio A decreasing values can also occur. Considering the buffer B one observes that even negative account values are possible. The policy finally seems to be less risky than the corresponding asset portfolio.

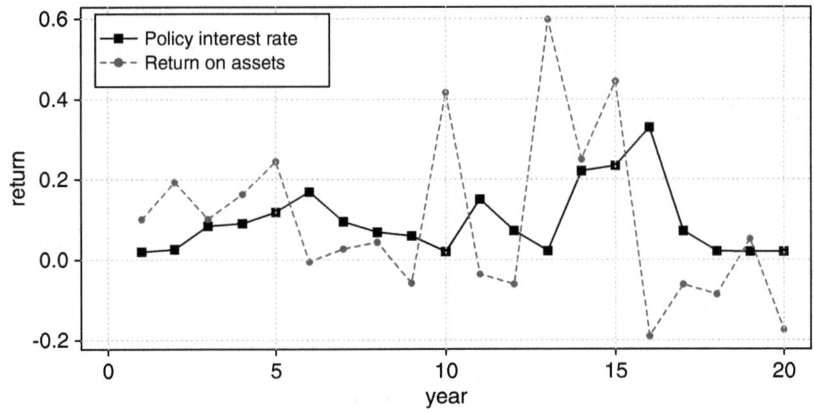

Fig. 2.2 A sample simulation of the policy interest rate $r_P(t)$ and the return on the reference portfolio $A(t)$ with parameters $r_G = 0.02, \mu = 0.05, \sigma = 0.2, \alpha = 0.3, \beta = 0.1, P_0 = 100, B_0 = 10$

Based on the model introduced above the authors organise their valuation of specific but artificial policies. In the concluding section they emphasise once more that life insurance companies have not paid serious attention to a proper valuation of

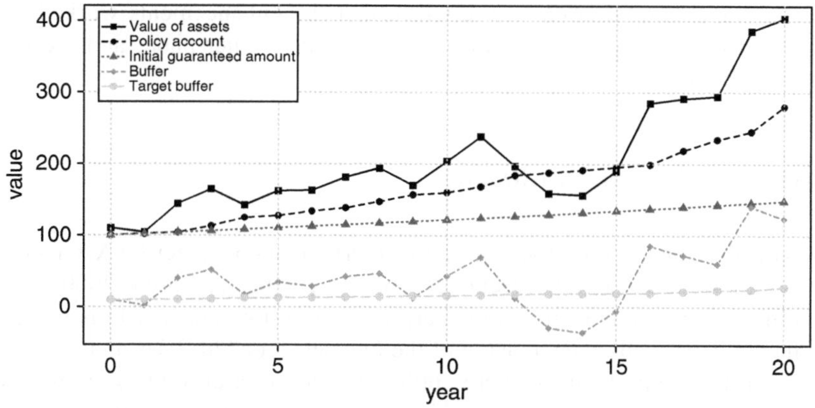

Fig. 2.3 A sample simulation of the asset value $A(t)$, the policy account $P(t)$, the initial guaranteed amount $P_0 \cdot (1 + r_G)^t$, the buffer $B(t)$ and the target buffer $\beta \cdot P(t)$ with parameters $r_G = 0.02, \mu = 0.05, \sigma = 0.2, \alpha = 0.3, \beta = 0.1, P_0 = 100, B_0 = 10$

embedded options in the past. It is shown that their modelled life insurance policy can be decomposed into three basic elements: A risk free bond, a dividend option and a surrender option.

To explore the properties of the model a quantitative analysis is presented in Grosen and Jørgensen (2000). They observe that the assumed bonus policy and the spread between the assumed expected return on assets μ and the interest rate guarantee r_G are key policy value drivers. Furthermore they show that in case of realistic parameter choice default probabilities of substantial size result.

An outlook for further research suggests applying *market value accounting* regarding the balance sheet as well as to the bonus policy. Additionally, they give examples on how to specify the model more realistically. One could take into account common features like periodic premiums, mortality, lapses and various expenses.

Jensen et al. (2001)

The paper Jensen et al. (2001) is co-written by P. L. Jørgensen and A. Grosen and basically an extension of their earlier published paper Grosen and Jørgensen (2000). The model developed therein shall be their starting point.

Mainly, the current paper presents two improvements. On the one hand they enhance the numerical procedure to calculate a policy value by reducing the dimensionality of the underlying algorithm. The authors denote that issue as the core contribution. On the other hand the former model becomes more realistic as mortality is taken into account.

In the concluding section they confirm that the enhancements worked out in the current paper support the results created in the former paper and are indeed an improvement.

Grosen and Jørgensen (2002a)

This is the third and last pure academic paper written or co-written be A. Grosen and P. L. Jørgensen which I will present here. In Grosen and Jørgensen (2000) a multi-period model providing a smoothing surplus distribution mechanism is presented. This model is basically an extension of the one developed in Briys and de Varenne (1997). Further improvements were applied in Jensen et al. (2001) by enhancing some numerical procedures and by taking mortality into account to give the model slightly more realism.

Now the current paper makes use of formerly gained insights and improves the modelling approach by E. Bryis and F. de Varenne again by focussing on the market valuation of both, the equity and the liabilities. Additionally some kind of regulatory intervention mechanism is implemented in the model. The model basics were outlined when discussing Briys and de Varenne (1997) on pages 23ff. To describe the extensions presented in Grosen and Jørgensen (2002a) a lot of mathematical theory is necessary. Since this lies beyond the scope of that thesis I refer the interested reader, therefore, to the original paper.

There is a detailed discussion on where to use this valuation framework whereas two issues are highlighted. As one possible area of application the determination of model parameters characterising initially fair premiums is suggested. Another application possibility is the fair market valuation of a company's equity and liabilities after the inception of the policies taking into account changes in market conditions.

In the concluding section the authors make mention of one important issue for further research. So far their models always considered only one homogenous class of policies with one fixed interest rate guaranteed, paid for a single premium and neglecting mortality.[10] To get a real practical model more flexibility towards contractual specifications is absolutely necessary.

Grosen and Jørgensen (2002b)

This paper is an attempt by A. Grosen and P. L. Jørgensen to apply their earlier developed model on real existing data from a handful of Danish pension and life insurance (P&L) companies.[11] Specifically, its objective is to *test empirically*

[10]Mortality is considered in Jensen et al. (2001).

[11]Confer Grosen and Jørgensen (2000).

whether (the) proposed model is a good description of the actual interest rate crediting behavior of the six largest Danish P&L companies during the period 1991–2000.[12]

These six P&L companies are AP, Codan, Danica, PFA, Topdanmark and Tryg. Since these companies do not provide detailed information on bonus distribution decisions the authors serve themselves with publicly available data consisting of policy interest rates and solvency ratios from the sample insurance companies. Their data set covers a 10 year period from 1991–2000. When the paper was written the six sample insurance companies represented around 55% of the Danish P&L savings market.

I have introduced the underlying model when presenting Grosen and Jørgensen (2000) on pages 26 ff. To ease further reading I will recall the relevant variables. $r_P(t)$ denotes the time t policy interest rate consisting of a guaranteed rate r_G and a non-guaranteed, variable bonus interest rate r_B that is greater or equal zero. The policy interest rate is the interest rate actually credited to the policyholder. $B(t)$ and $P(t)$ stand for the time t value of the buffer and the policy reserve respectively. The ratio B/P shall be the measure for the solvency ratio as mentioned above.

Given the base formula for the policy interest rate one observes that r_P is specified by two variables, the distribution ratio α and the target buffer ratio β.[13] To describe their statistical model a slightly modified version of the base formula is taken. Define $\psi = \beta + r_G/\alpha$, then

$$r_P^i(t) = r_G + \max\left\{0, \alpha_1^i \frac{B^i(t-1)}{P^i(t-1)} - \alpha_1^i \psi^i\right\}, \tag{2.14}$$

where the subscript i, $i \in \{1, 2, \ldots 6\}$ indexes the different companies considered. Note that for simplifying matters r_G is assumed to be fixed at 4.5% by the authors throughout their analysis.[14] As explained the available data consists of a time series of the policy interest rate $r_P^i(t)$ and the relation $\frac{B^i(t-1)}{P^i(t-1)}$ for each i. With

$$y_t^i = r_P^i(t) - r_G, \tag{2.15}$$

$$x_t^i = \frac{B^i(t-1)}{P^i(t-1)} \tag{2.16}$$

and

$$\alpha_0^i = -\alpha_1^i \psi^i, \tag{2.17}$$

[12] See Grosen and Jørgensen (2002b).

[13] $r_P(t) = \max\left\{r_G, \alpha \cdot \left(\frac{B(t-1)}{P(t-1)} - \beta\right)\right\}$

[14] See the respective explanations in Grosen and Jørgensen (2002b).

their statistical model corresponding to Eq. 2.14 is given by the following expression.

$$y_t^i = \max\left\{0, \alpha_0^i + \alpha_1^i x_t^i + \epsilon_t^i\right\}, \qquad \epsilon_t^i \sim \mathcal{N}(0, \sigma_i^2) \text{ i.i.d} \tag{2.18}$$

To estimate the parameters α_0^i, α_1^i and σ_i^2 a maximum likelihood (ML) approach of Amemiya (1973) is used. To give the reader an impression of the results the average values for the ML estimators are plotted in Table 2.3. For a more detailed reporting I refer to the original paper.

Table 2.3 Average values of the estimated parameters $\hat{\alpha}^i$ and $\hat{\Psi}^i$ for the Grosen and Jørgensen model (Source: Grosen and Jørgensen (2002b))

Average ML estimators		Average fit measure
$\hat{\alpha}$	$\hat{\Psi}$	R^2
0.1686	0.0670	0.455

The authors conclude that, even though their sample was rather poor, the model enables a good fit of the observed data. Regarding the definition of bonus policy in practice they state to have gained some valuable insights from the results. Certainly this model may be of interest for other markets than the Danish.

Further improvements are suggested by additional explanatory variables like mortality, surrender or other parameters affecting the determination of premiums. A short but interesting comment is the author's forecast that moving forward to *market value based valuation* will become a standard and also mean a great challenge to the P&L companies. As we know today this has become perfectly true.

A. Grosen and P. L. Jørgensen close their paper by mentioning that the estimated bonus mechanism parameters could also be viewed as a ranking and comparison measure for particular insurance policies.

Hansen and Miltersen (2002)

The authors of Hansen and Miltersen (2002) present a further modelling framework to analyse minimum interest rate guarantee and dividend option. For modelling purposes they propose to make use of the smoothing surplus distribution mechanism used by most Danish life insurance companies and pension companies, basically an extension of the model presented in Grosen and Jørgensen (2000).

Their model considers a pure endowment insurance policy containing a dividend option. Bonuses are assumed to be allocated on a yearly basis from some bonus reserve and finally paid out at maturity T. On top of, that the policyholder might receive a terminal bonus if the bonus reserve is positive at the end of its contract. If the bonus reserve is negative then, the deficit will be covered by the company. De facto this is equivalent to issuing a put option on the bonus reserve.

Within their paper two different cases are considered. Firstly, there is only one policyholder. Here the authors characterise fair contracts between the policyholder and the company. Secondly, they investigate two policyholders that can differ with respect to the minimum interest rate guarantee, entry dates or exit dates. For this setting different distribution mechanisms for the terminal bonus are proposed.

For the one policyholder case their model is based on a simplified balance sheet approach similar to the approaches suggested by Briys and de Varenne (1997) or Grosen and Jørgensen (2000). On the asset side, X denotes the company's total assets. Initially the policyholder pays that amount X and the company invests it in some reference portfolio. On the liability side, there are the *policyholder's account* A, the *bonus reserve* B and the *company's account* C. See Table 2.4 therefore:

Table 2.4 The company's initial balance sheet in the Hansen and Miltersen model (Source: Hansen and Miltersen (2002))

Assets	Liabilities
X	A
	B
	C
X	X

At start the initial deposit paid by the policyholder is credited to the policyholder's account A, hence $A(0) = X(0) = X$. To model the reference portfolio X a geometric Brownian motion is used. The time t value of the reference portfolio, $X(t)$ is then described by the following stochastic differential equation.

$$dX(t) = rX(t)dt + \sigma X(t)dW(t), \qquad X(0) = X, \tag{2.19}$$

under the equivalent martingale measure Q. r and σ denote the instantaneous risk free interest rate and the instantaneous volatility of the reference portfolio respectively. W is a standard Wiener process under Q. For further information on equivalent martingale measures the interested reader is referred to Harrison and Kreps (1979) and Harrison and Pliska (1981).

Due to the interest rate guarantee at least an interest rate of g is credited to the policyholder's account A at the end of each year. To understand the bonus distribution mechanism it is easiest to describe first the bonus allocation to both, the policyholder's account A and the company's account C. In the latter the company collects payments for issuing and guaranteeing contractual benefits.

Primarily depending on the ratio $B/(A+C)$ a possible extra amount on top of the minimum interest rate guaranteed g is determined. So the sum $A(t-1) + C(t-1)$ shall be continuously compounded at the rate $r_{AC}(t)$, where

$$r_{AC}(t) = \max\left\{g, \ln\left(1 + (\alpha + \rho)\left(\frac{B(t-1)}{A(t-1) + C(t-1)} - \gamma\right)\right)\right\}, \tag{2.20}$$

with $\alpha + \rho \in [0,1]$ and $t \in [1, 2, \ldots, T]$. The evolution of the sum $A + C$ is then given by

$$(A(t) + C(t)) = (A(t-1) + C(t-1))\, e^{\max\{g,\lambda\}}, \tag{2.21}$$

with $\lambda = \ln\left(1 + (\alpha + \rho)\left(\frac{B(t-1)}{A(t-1)+C(t-1)} - \gamma\right)\right)$, $\alpha, \rho \in [0,1]$ and $\alpha + \rho \in [0,1]$. To split the evolution of A and C a percentage payment fee $\xi \in [0,1]$ is introduced. Modifying Eq. 2.21 leads to

$$A(t) = A(t-1)e^{\max\left\{g,\ln\left(1+\alpha\left(\frac{B(t-1)}{A(t-1)+C(t-1)}-\gamma\right)\right)\right\}-\xi} \tag{2.22}$$

and

$$C(t) = (A(t) + C(t)) - A(t). \tag{2.23}$$

Considering Formula 2.21 and 2.23 we see two different ways the company can collect payment into its account C for issuing the minimum interest rate guarantee. Either we have $\xi > 0$ and $\rho = 0$, called the *direct* method for collecting payments, or the opposite case with $\rho > 0$ and $\xi = 0$, called the *indirect* method for collecting payments. In the subsequent the authors focus on the direct method.

The bonus reserve B is a residual quantity and evolves according to the following equation. With $B(0) = B = 0$ we have

$$\begin{aligned} B(t) &= B(t-1) + X(t) - X(t-1) - (A(t) + C(t)) \\ &\quad + (A(t-1) + C(t-1)). \end{aligned} \tag{2.24}$$

At maturity date T the policyholder receives the amount of its account $A(T)$ plus – if positive – the terminal value of the bonus account $B(T)$. In the opposite case where $B(T)$ is negative the company covers the deficit with its account $C(T)$.

To determine a fair set of contract parameters like g, α, ρ, ξ and γ, M. Hansen and K. R. Miltersen define a fairness condition. Let

$$V_t(Z(T)) = e^{-r(T-t)} E_t^Q[Z(T)] \tag{2.25}$$

be the time t *market value operator* with $E_t^Q[\cdot]$ the conditional expectation under the equivalent martingale measure Q and $Z(T)$ some stochastic payoff at date T. Then the fairness condition is given by

$$X = V_0(A(T) + B^+(T)) \Leftrightarrow 1 = V_0\left(\frac{A(T)}{X}\right) + V_0\left(\frac{B^+(T)}{X}\right), \tag{2.26}$$

where B^+ denotes positive values of the bonus account.

I wish to refer to the second purpose of this paper, investigating the two policyholder case. Modelling two (or several) policyholders instead of just one means deciding whether to pool bonus reserves or not. Using the same framework as introduced above, the authors analyse various contractual specifications for two policyholders and terminal bonus distribution schemes with respect to pooled and non-pooled bonus reserves.

The authors conclude that their model is able to price life insurance contracts containing a minimum interest rate guarantee and a dividend option using the principle of fair valuation. Due to this there is no need for an up-front premium. The policyholder pays for the guarantee via the annual payment fee (ξ).

Analysing the two policyholder case shows that when pooling inhomogeneous policyholders, it is more favourable from the company's point of view to leave at least one of the policyholders worse off. Furthermore the modelling results indicate that old policyholders with a higher interest rate guarantee "cheat" newer policyholders with a lower guarantee with respect to a fair distribution of the pooled bonuses.

M. Hansen and K. R. Miltersen finalise with the remark that life insurance companies should not pool the policyholder's bonus reserve, implying the calculation of an individual payment fee for each policyholder.

Haberman et al. (2003)

This paper is motivated by a handful of reasons for public concern about the financial health of life insurance companies. Whilst former presented papers identified fallen interest rates, fallen equity markets and deliberate mismatching in favour of equities as the main risk drivers, Haberman et al. (2003) mention two new and important issues on that matter: The move towards more transparency with the customers and the international move towards market based accountancy standards.

Based on these insights the purpose of the current paper is to develop a suitable valuation framework for participating life insurance policies. Conceptually that valuation framework is derived from Grosen and Jørgensen (2000). Again they assume the policy reserve P evolves according to the following formula.

$$P(t) = (1 + r_P(t)) \cdot P(t-1), \qquad t \in I, \tag{2.27}$$

where $I = \{1, 2, \ldots, T\}$. Instead of just one definition for the policy interest rate r_P three different smoothing schemes are suggested to define r_P.[15] Note that these schemes are not imaginary but based on evidence from Needleman and Roff (1995), Chadburn (1998) and Tillinghast (2001).

[15]Confer Formula 2.11.

Scheme I

Let $\beta \in (0, 1)$ be the *participating rate* and $n = \min(t, \tau), \tau < T$, the length of the smoothing period. Then

$$r_P(t) = \max\left\{r_G, \frac{\beta}{n}\left(\frac{A(t)}{A(t-1)} + \ldots + \frac{A(t-n+1)}{A(t-n)} - n\right)\right\}. \quad (2.28)$$

Scheme II

Let β and n be defined as for scheme I. Then

$$r_P(t) = \max\left\{r_G, \beta\left(\sqrt[n]{\frac{A(t)}{A(t-n)}} - 1\right)\right\}. \quad (2.29)$$

Scheme III

Let P^1 be the unsmoothed asset share defined as

$$P^1(t) = P^1(t-1)(1 + r_P(t)) \quad (2.30)$$

and

$$r_P(t) = \max\left\{r_G, \beta\frac{A(t) - A(t-1)}{A(t-1)}\right\}. \quad (2.31)$$

Then the policy reserve P is defined as follows

$$P(t) = \alpha P^1(t) + (1-\alpha)P(t-1), \quad (2.32)$$

with $\alpha \in (0, 1)$ playing the role of the smoothing parameter.

At maturity date T the policyholder receives the terminal value of its policy reserve $P(T)$ plus a fraction $\gamma R(T)$, $\gamma \in (0, 1)$ of a possible surplus generated by the reference portfolio A denoted by

$$R(T) = \max(0, A(T) - P(T)). \quad (2.33)$$

On the other side the insurance company gets what is left. This can be either a gain if $A(T) \geq P(T)$ or a loss if $A(T) < P(T)$. A potential loss shall be denoted by

$$D(T) = \max(0, P(T) - A(T)). \quad (2.34)$$

S. Haberman, L. Ballotta and N. Wang conclude with a numerical analysis of the three different smoothing schemes. Using the risk neutral valuation they focus on the following quantities.

$$V_P(0) = E^Q[e^{-rT} P(T)], \tag{2.35}$$

$$V_R(0) = E^Q[e^{-rT} R(T)] \tag{2.36}$$

and

$$V_D(0) = E^Q[e^{-rT} D(T)], \tag{2.37}$$

with r the instantaneous risk free interest rate. By using different smoothing schemes, the model becomes slightly more realistic and enables a comparison in this regard. But since their analysis is based on artificial life insurance contracts, only the numerical results do not really provide new and valuable insights compared to earlier works.

Cummins et al. (2004)

This paper investigates various implementations of interest rate guarantees and dividend options across different countries. The authors focus on modelling common practice in Denmark, Germany, Norway, the United Kingdom and the United States.

The basic modelling set-up formulated in Cummins et al. (2004) is comparable to the balance sheet approaches presented in Grosen and Jørgensen (2000) or in Hansen and Miltersen (2002), whereas the surplus distribution schemes naturally differ from country to country. Insurance specific factors like mortality or periodic premiums are as usually neglected. Furthermore only one asset investment opportunity, called the *market index*, is taken into account.

J. D. Cummins, K. R. Miltersen and S.-A. Persson remark that in European contracts the surplus distribution mechanism is usually designed more sophisticated than in the US equivalent where, for example, no bonus account is modelled. However, given the objective is to provide the policyholder a future benefit with low risk, their numerical analysis indicates that United States' policies outperform European policies.

They further observe that the Danish, German and the United Kingdom contracts including more sophisticated smoothing mechanisms basically do not perform significantly differently than the assumed market index. This means that the minimum interest rate guarantee and the dividend option have virtually no valuable impact on the policyholder's benefits.

Kling et al. (2007)

In the papers discussed so far various ways to give interest rate guarantees are considered. Remember the point-to-point scheme suggested in Briys and de Varenne (1997). Hence the guarantee is only relevant at maturity. As opposed to this we know year-by-year (or cliquet-style) guarantees as presented e.g. in Grosen and Jørgensen (2000) or Hansen and Miltersen (2002), where the interest rate guarantee is credited yearly. I have further presented papers with different mechanisms *defining* a surplus amount. Therefore see Haberman et al. (2003), for example.

Lastly Kling et al. (2007) consider mechanisms *distributing* annual surpluses to policyholders. For their purposes they present three different possibilities.

1. Surpluses may be allocated to the policy account. Thus allocated surpluses are charged with the same interest rate as the policy account.
2. Surpluses may be allocated to a separate account. That accountÊis owned by the policyholder but there is no interest credited on it.
3. Surplus may be allocated to a terminal bonus account. The policyholder will receive a possible terminal bonus only at maturity or, depending on contractual specifications, a reduced amount thereof when surrendering. As long as the policy is in force, the insurance company may use that account to pay the minimum interest rate guarantee if the return on their assets does not yield the minimum interest rate guaranteed in any particular year.

Thus, the purpose of the current paper is to analyse numerical impacts on the risk exposure of a life insurance company due the different surplus distribution mechanisms presented above. For this the authors use their own modelling framework based on a balance sheet approach enabling the consideration of year-by-year as well as point-to-point guarantees. The focus is, however, on the different surplus distribution mechanisms.

The numerical analysis shows clearly that the third distribution mechanism significantly differs from mechanism one and two. The third mechanism faces a much lower shortfall risk than the other two. Thus, a company applying that mechanism may invest its funds in riskier asset classes promising higher expected returns and also provide a higher interest rate guarantee for this.

A. Kling, A. Richter and J. Russ notice that the third mechanism is severely restricted by regulators in some European countries whilst it is not in some other European countries. This may lead to a distortion of competition since the current analysis has indicated some strong pros of this distribution mechanism. They conclude that severe regulations are not always advantageous for the insurance industry and could even be counterproductive as shown by this example.

2.3 Conclusions for the Modelling Framework

In this chapter I have defined basic types of life insurance products and introduced common options and guarantees embedded therein. I have further appointed the most relevant ones for the purpose of this thesis: Minimum interest rate guarantee and dividend option. Based on that criterion, a literature survey outlining the reader the actual state of the scientific knowledge was presented.

Remember the objectives formulated in Sect. 1.3 where I have requested a quantitative modelling approach to handle embedded options and guarantees based on modern valuation techniques with respect to current regulatory and accounting standards. Taking the literature survey we have seen several different approaches as to how such a model could be designed. But in order to achieve these objectives I definitely identify some important issues that have not or have just insufficiently been taken into account so far. Thus to finalise the current chapter, I aim to outline theoretical characteristics for the modelling framework I will formulate in the upcoming chapter. Hereby, certain features from existing works are adopted and my own extensions are presented.

- As previously announced I basically aim to design a fairly general and transparent framework that provides an appropriate substructure to handle every kind of embedded option and guarantee. In the end the focus will be on minimum interest rate guarantee and dividend option and so the framework finally being presented and applied for modelling purposes in the subsequent chapters will be some special case thereof. But note that this does not affect the requested flexibility of the model itself. Having said this, a year-by-year balance sheet approach as suggested by Grosen and Jørgensen (2000) seems to be a good starting point. Take notice of the fact that this model is more than just a theoretical construct but also widely accepted in practice.[16]
- According to Sect. 1.3 the central question I aim to answer with that model is about the value the considered embedded options and guarantees do have, absolutely and relative to the current policy reserves. Also remember that market consistency is the basic prerequisite for valuation methodologies applied nowadays and so this model must be set up. For this reason it shall be calibrated consistently to true market data on the asset side and allow for considering real, existing and in-force policy data on the liability side. Except Grosen and Jørgensen (2002b), discussed papers in the literature survey consequently set up their analysis on artificial insurance policies decreasing practical benefits from the respective conclusions.
- Referring to the selected balance sheet approach, the asset side shall be modelled stochastically. As stated above the underlying stochastic process will be calibrated to true market data. But with a simple Brownian motion, as applied in most of the papers that have been discussed in the literature survey,

[16]Confer Hansen and Miltersen (2002).

one does not achieve an accurate mapping of an insurance company's asset portfolio. Therefore, considering the asset portfolio of a real existing insurance company leads to a joint stochastic modelling of three asset types: Two correlated processes for an equity and a property price index and *zero coupon bond* (ZCB) prices.

- Most of the innovation will happen on the liability side. First of all, a set of approximately 110,000 real existing life insurance policies will provide the data basis for the model. Since there is no market for these in-force policies a market consistent value needs to be determined. Hence mortality and expenses are to be taken into account, too. Note that the only paper partially considering mortality was Jensen et al. (2001). Explicitly mentioning the importance of market consistent valuation of both, the equity and the liabilities, could be found in Grosen and Jørgensen (2002a) only.
- Haberman et al. (2003) addresses different mechanisms to determine a possible surplus. Indeed this has significant impacts on the entire model in general and on the dividend option in particular. Thus, I will make allowance for handling that issue in detail.
- With Kling et al. (2007) I have discussed one paper investigating different mechanisms to distribute a possible surplus, another crucial element regarding the value of the dividend option. Since that distribution mechanism is contractually specified for the given policy data, only performing a "what-if analysis" makes sense.
- A last point worth noting here concerns a company's shareholders. Even though the model by Briys and de Varenne (1997) incorporates some equity component, the shareholders interests are not accurately incorporated there in my opinion.

In the following Chaps. 3 and 4 I will formulate the entire modelling framework based on the theoretical outline given above and present all market and insurance specific data needed therefore.

Chapter 3
Asset Modelling Process

In this chapter I will introduce the first part of the entire model framework, the asset modelling process. The point to start is given by a year-by-year balance sheet approach as presented in Grosen and Jørgensen (2000). For the purpose of this thesis some extensions are in order. Let the following table denote the model's underlying balance sheet.

Table 3.1 The underlying balance sheet of the model framework at time t

Assets		Liabilities	
Assets portfolio	$A(t)$	Policy account	$P(t)$
		Buffer	$B(t)$
		Undistributed surplus	$U(t)$
Total	$A(t)$	Total	$A(t)$

The intention of this and the subsequent chapter's sections is twofold. On the one hand, I will describe the evolution of the respective balance sheet entries over time. The current chapter addresses the asset modelling, the subsequent chapter addresses the liability modelling. On the other hand, I will define the initial values for each of the balance sheet entries. To define the initial value of the asset portfolio, one needs to know the idea of the whole modelling approach, which means after having introduced the liability modelling. Furthermore it is also necessary to outline the available data that is needed for calibrating and calculation purposes. By this, the entire model framework is defined.

A general remark regarding the data used. Mainly for calibration reasons, financial market data will be used. For this publicly available data is taken whenever possible. Otherwise I am going to access to historical market data via Datastream.[1]

[1] *"Thomson Reuters Datastream is a global research solution that enables sophisticated time series analysis across a broad range of financial instruments."* For further information see http://thomsonreuters.com/content/financial/pdf/i_and_a/datastream_corporate.pdf. Last checked: 1 June 2011.

N. Rüfenacht, *Implicit Embedded Options in Life Insurance Contracts*, Contributions to Management Science, DOI 10.1007/978-3-7908-2843-6_3,
© Springer-Verlag Berlin Heidelberg 2012

Insurance specific data is kindly provided by *Helvetia Versicherungen*, called Helvetia henceforth.[2] Furthermore all data are collected as at the end of 2009.

All calculations including the calibration procedures of the particular models and the simulation routines of the entire model will be performed in **R**, a free open-source and object orientated programming language. For further information, a variety of introductory literature or a full language documentation browse the official **R** webpage.[3]

3.1 Defining a Proxy for the Asset Modelling

Referring to the balance sheet in Table 3.1, its assets shall be modelled according to some stochastic process. To determine a feasible proxy for that process I will analyse Helvetia's entire life insurance asset portfolio first. The reader can find a public and detailed view on it on the FINMA webpage.[4] In Table 3.2 there is a slightly abridged listing of these assets for the purpose of this thesis and thus this is not a unique listing.

Table 3.2 Helvetia's summarised life insurance asset portfolio as at the end of 2009. Amounts of money are denoted in Swiss Francs (Source: Rüfenacht/FINMA)

Asset	Amount	Ratio(%)
Cash	803,914,199	4.0
Shares	913,516,683	4.5
Property	3,250,622,862	16.0
Bonds	11,177,042,694	55.0
Mortgages	2,974,165,204	14.6
Else	1,197,990,731	5.9
Total	20,317,252,373	100

This summarised asset portfolio is a first step towards the requested proxy. Given that summarised asset portfolio, three asset classes are derived. Each class shall contain homogenous asset types in a sense that they can be modelled according to the same stochastic process. The classes are represented by shares, property and bonds. The idea behind this is to define a stochastic process for each class. For shares and property two correlated geometric Brownian motions will be applied, to model bonds I will make use of an interest rate model.

In the subsequent, shares and property are modelled with a given correlation but independently of bonds. Indeed this is a very strong argument, but realise that correlation between stock markets and bond markets is known to be highly volatile

[2]Helvetia Schweizerische Lebensversicherungsgesellschaft AG, Basel.

[3]See http://www.r-project.org/. Last checked: 1 June 2011.

[4]See http://www.versichererreport.finma.ch/reportportal/. Last checked: 1 June 2011.

over time.[5] So on the one hand assuming some constant correlation for modelling purposes would not be plausible and, on the other hand, modelling a time varying correlation would present a delicate issue. Hence assuming non-correlated processes is a legitimate approach in my opinion.

However, before introducing the particular stochastic processes, see in Table 3.3 the three asset classes assumed. Again this is neither an official nor a unique listing but only a composition to define a possible modelling proxy. This is also the reason why the residual and unspecified position "Else" in Table 3.2 is left away in the subsequent.

Table 3.3 Selected asset classes derived from the summarised balance sheet of Helvetia's life insurance business as at the end of 2009. Amounts of money are denoted in Swiss Francs

Class	Contents	Amount	Ratio(%)
Shares	Shares	913,516,683	4.8
Property	Property	3,250,622,862	17.0
ZCB prices	Cash, bonds, mortgages	14,955,122,097	78.2
Total		19,119,261,642	100

The geometric Brownian motion is a well known and widely accepted approach to model, for example, shares stochastically. To be more precise, it is the price of the share that will be modelled. Furthermore, I assume property to behave similar to shares, thus the prices of this asset class will be modelled according to a geometric Brownian motion, too. Following Shreve (2004) I will provide an overview on the theoretical background of the geometric Brownian motion in the subsequent section. A more detailed mathematical treatment lies beyond the scope of this thesis and therefore the interested reader is referred to the cited literature.

3.2 Basics of Stochastic Modelling

Assume a fair coin. Let p and $q = 1 - p$ be the probability of H (head) and T (tail) on each coin toss. Since the coin is assumed to be fair, we must have $p = q = 0.5$. By repeatedly tossing this coin one can construct a *symmetric random walk.*

Definition 4 (symmetric random walk). Let ω_n be the outcome of the nth toss of a fair coin. Define

$$X_j = \begin{cases} 1 \text{ if } \omega_j = H, \\ -1 \text{ if } \omega_j = T, \end{cases} \tag{3.1}$$

[5]For further information on this topic the interested reader is referred to Barsky (1989) and Stivers and Sun (2002).

and

$$M_k = \sum_{j=1}^{k} X_j, \qquad k = 1, 2, \ldots \text{ and } M_0 = 0. \tag{3.2}$$

Then the process $M_k, k = 0, 1, 2, \ldots$ defines a *symmetric random walk.*

In Fig. 3.1 ten steps of a sample symmetric random walk are plotted.

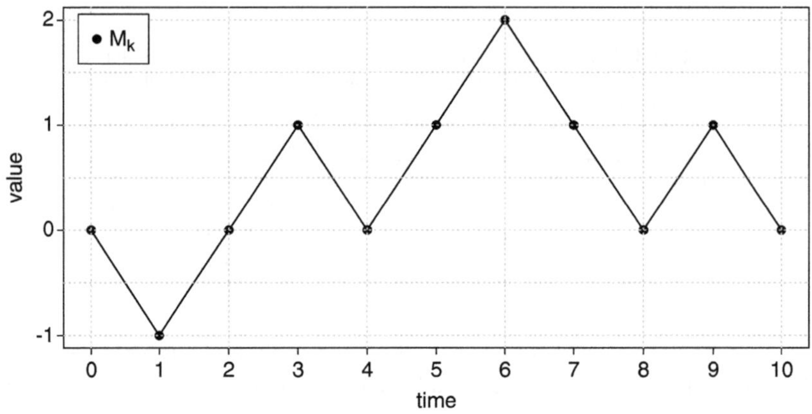

Fig. 3.1 Ten steps of a sample symmetric random walk M_k

The increments of a symmetric random walk are independent. This means that, given a series of nonnegative integers $0 = k_0 < k_1 < \ldots < k_m$, the random variables

$$M_{k_1} - M_{k_0}, M_{k_2} - M_{k_1}, \ldots, M_{k_m} - M_{k_{m-1}}$$

are independent. Each of these random variables

$$M_{k_{i+1}} - M_{k_i} = \sum_{j=k_i+1}^{k_{i+1}} X_j$$

is called an *increment* of the symmetric random walk. Moreover each increment $M_{k_{i+1}} - M_{k_i}$ has an expected value of 0 and a variance of $k_{i+1} - k_i$. To prove this we use $\text{Var}(X_j) = \mathbb{E}[X_j^2] - \mathbb{E}[X_j]^2 = 1 - 0 = 1$, and since the different X_j are independent we have

$$\text{Var}(M_{k_{i+1}} - M_{k_i}) = \sum_{j=k_i+1}^{k_{i+1}} \text{Var}(X_j) = \sum_{j=k_i+1}^{k_{i+1}} 1 = k_{i+1} - k_i. \tag{3.3}$$

Now we can extend the symmetric random walk by speeding up time and scaling down the step size. This leads to the definition of the *scaled symmetric random walk.*

Definition 5 (scaled symmetric random walk). Let n and t be some positive integer and M_{nt} a symmetric random walk. Then

$$W^{(n)}(t) = \frac{1}{\sqrt{n}} M_{nt} \tag{3.4}$$

defines a *scaled symmetric random walk.*

In Fig. 3.2 a sample path of a scaled symmetric random walk $W^{(100)}$ is plotted.

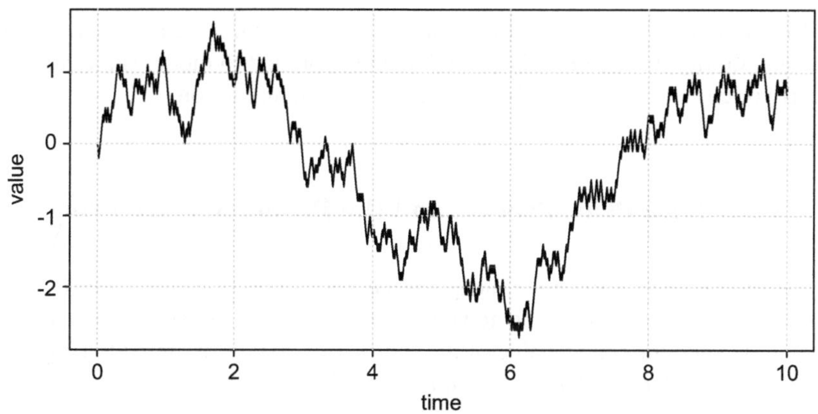

Fig. 3.2 A sample path of a scaled symmetric random walk $W^{(100)}$

Like the symmetric random walk M_k also the scaled symmetric random walk $W^{(n)}$ has independent increments. Let $0 = t_0 < t_1 < \ldots < t_m$ such that each nt_j is an integer. Then the increments

$$W^{(n)}(t_1) - W^{(n)}(t_0), W^{(n)}(t_2) - W^{(n)}(t_1), \ldots, W^{(n)}(t_m) - W^{(n)}(t_{m-1})$$

are independent. Furthermore one can show that

$$\mathbb{E}[W^{(n)}(t) - W^{(n)}(s)] = 0 \tag{3.5}$$

and

$$\mathrm{Var}(W^{(n)}(t) - W^{(n)}(s)) = t - s \tag{3.6}$$

with $0 \leq s \leq t$, such that ns and nt are integers. Finally the Brownian motion is obtained from the scaled symmetric random walk $W^{(n)}$ as $n \to \infty$. Its proper mathematical definition is as follows.

Definition 6 (Brownian motion). Let $(\Omega, \mathcal{F}, \mathbb{P})$ be a probability space. For each $\omega \in \Omega$, suppose there is a continuous function $W(t), t \geq 0$ that satisfies $W(0) = 0$ and that depends on ω. Then $W(t), t \geq 0$ is a *Brownian motion* if for all $0 = t_0 < t_1 < \ldots < t_m$ the increments

$$W(t_1) - W(t_0), W(t_2) - W(t_1), \ldots, W(t_m) - W(t_{m-1})$$

are independent and each of these increments is normally distributed with

$$\mathbb{E}[W(t_{i+1}) - W(t_i)] = 0 \tag{3.7}$$

$$\text{Var}[W(t_{i+1}) - W(t_i)] = t_{i+1} - t_i \tag{3.8}$$

The Brownian motion is also called a *standard Wiener process*. Having defined the Brownian motion we can finally formulate the approach to model share prices and property prices stochastically: The geometric Brownian motion.

3.3 The Share Price Process and the Property Price Process

Let us suppose $S_0 = 1$ to be the today's price of one unit of some share. This share shall have a constant yearly expected rate of return μ and a constant yearly volatility rate $\sigma > 0$. Then the time t value of the share price $S(t)$ is assumed to be given by the following *stochastic differential equation* (SDE).

$$dS(t) = \mu S(t)dt + \sigma S(t)dW(t), \qquad S(0) = S_0, \tag{3.9}$$

where $W(t)$ denotes a Brownian motion. Equation 3.9 defines a *geometric Brownian motion* providing an appropriate approach to model the prices of shares or property.

Figure 3.3 shows a sample path of a geometric Brownian motion S with parameters $\mu = 0.2$ and $\sigma = 0.5$.

Taking Eq. 3.9 one can show that the relative change in $S(t)$ is lognormal distributed as

$$\log \frac{S(t)}{S(0)} \sim \Phi\left((\mu - \frac{\sigma^2}{2})t, \sigma\sqrt{t}\right), \tag{3.10}$$

where log denotes the natural logarithm.

A Brownian motion and hence also a geometric Brownian motion is a continuous processes that cannot exactly be mapped by a computer. Hence the processes need to be approximated by a discretisation. Since the increments of a Brownian motion are normally distributed, a geometric Brownian motion can be approximated as shown in the following **R** source code segment.

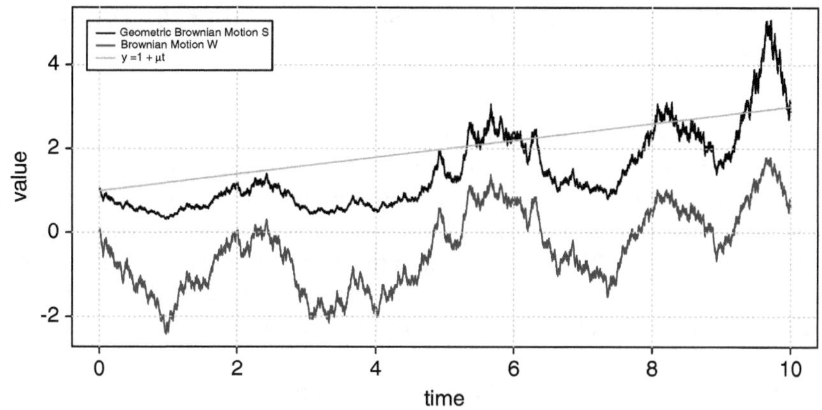

Fig. 3.3 Sample path of a geometric Brownian motion S with parameters $\mu = 0.2$ and $\sigma = 0.5$

```
S[t+1] <- S[t] + mu*S[t]*dt
          + sigma*S[t]*rnorm(1,0,1)*sqrt(dt)
```

We now assume both, shares and property to behave similarly but not equally. This means that we need two correlated stochastic processes to be modelled simultaneously. Let $S(t)$ and $P(t)$ be the time t price of one unit of shares and property respectively. Then

$$\begin{aligned} dS(t) &= \alpha S(t)dt + \sigma S(t)dW^1(t), \quad S_0 = 1, \\ dP(t) &= \beta P(t)dt + \eta P(t)dW^2(t), \quad P_0 = 1, \end{aligned} \tag{3.11}$$

where (W^1, W^2) is a two-dimensional Brownian motion with instantaneous correlation ρ as from

$$dW^1(t)dW^2(t) = \rho dt, \tag{3.12}$$

and positive constants $\alpha, \beta, \sigma, \eta$ and $-1 \leq \rho \leq 1$. With the Eq. 3.11 the stochastic processes for modelling the first two classes in Table 3.3, shares and property, are given. An illustrative example of modelled shares and property is presented in the upcoming section. In order to complete the processes need to be calibrated.

3.3.1 *Calibration of the Share Price Process and the Property Price Process*

Recall the stochastic process for the share price S in Eq. 3.11. As a next step the model parameters α and σ need to be estimated. This asks for a reasonable data basis. Helvetia's asset portfolio consists of shares and similar assets approximately matching the structure of the *Swiss Performance Index* ® (SPI). Thus, I will take historic data from the *SPI total return index* that can be found on the SIX webpage.[6] Unlike a price index a total return index assumes dividends and other payouts to be reinvested immediately. This is exactly the way I am arguing for modelling purposes.

Let SPI_t be the closing value of the SPI in t where I will consider a 5 year period from 1.1.2005 to 31.12.2009 for t. $n+1 = 1{,}257$ shall be de number of data observed in this period. Using the lognormal property of the share price process given by Formula 3.10, $\log \frac{\mathrm{SPI}_t}{\mathrm{SPI}_{t-1}}$ is calculated for all t. Now the estimators for α and σ in Eq. 3.11 can easily be derived. Let $\bar{m}$ be the mean log return of the SPI in the period considered:

$$\bar{m} = \frac{1}{n}\sum_t \log\frac{\mathrm{SPI}_t}{\mathrm{SPI}_{t-1}} = 0.00022. \tag{3.13}$$

Then $\hat{\sigma}$ is given by

$$\hat{\sigma} = \sqrt{\frac{\frac{1}{n-1}\sum_t \left(\log\frac{\mathrm{SPI}_t}{\mathrm{SPI}_{t-1}} - \bar{m}\right)^2}{dt}} = 0.1964. \tag{3.14}$$

Since we are observing $n+1 = 1{,}257$ data points within 5 years I set $dt = 5/1{,}257 \approx 0.00398$.

Having once determined the estimator for σ we can continue by estimating $\bar{\alpha}$.

$$\bar{m} = \left(\hat{\alpha} - \frac{\hat{\sigma}^2}{2}\right)dt = 0.00022 \tag{3.15}$$

$$\Leftrightarrow \hat{\alpha} = \frac{\bar{m}}{dt} + \frac{\hat{\sigma}^2}{2} = 0.07357 \tag{3.16}$$

So what data basis shall be taken as a proxy for the property price process P? Indeed there are suitable real estate indices provided by the SIX group.[7] For our purposes

---

[6] Quotes available on http://www.six-swiss-exchange.com/indices/shares/spi_family_de.html. Last checked: 1 June 2011.

[7] Quotes available on http://www.iazicifi.ch/de/indizes.php. Last checked: 1 June 2011.

I am going to use the *SWX IAZI Investment Real Estate Performance Index* (IAZI index), a performance index that covers development of prices and net cash flow return on Swiss real estate. Note that this index is again a total return index. For both indices a chart of the data used for the calibration procedure can be seen on Fig. A.1 in Appendix A.

Analogously we can calculate estimators for the property price process P then. Taking into account that the IAZI index is available only quarterly we get

$$\hat{\eta} = 0.0444 \tag{3.17}$$

and

$$\hat{\beta} = 0.0781. \tag{3.18}$$

To calculate the correlation coefficient ρ as defined in Eq. 3.12, we need to pick quarterly SPI data in accordance with the IAZI data. Define with $T_Q = \{4Q2004, 1Q2005, \ldots, 4Q2009\}$ all quarter end time points between 1.1.2005 and 31.12.2009 assuming that $4Q2004$ is equal to 1 January 2005. With $\text{IAZI}_{t^*}, t^* \in T_Q$ denote the observed value for the IAZI index as at t^*. Now define

$$x_{t^*} = \log \frac{\text{SPI}_{t^*}}{\text{SPI}_{t^*-1}}, \qquad t^*, t^* - 1 \in T_Q, \tag{3.19}$$

$$y_{t^*} = \log \frac{\text{IAZI}_{t^*}}{\text{IAZI}_{t^*-1}}, \qquad t^*, t^* - 1 \in T_Q. \tag{3.20}$$

With $\bar{x}$ and $\bar{y}$ the mean values over all x_{t^*} and y_{t^*} respectively, the estimator $\hat{\rho}$ for the yearly correlation coefficient is given by

$$\hat{\rho} = \frac{\sum_t (x_t - \bar{x}) \cdot (y_t - \bar{y})}{\sqrt{\sum_t (x_t - \bar{x})^2 \cdot \sum_t (y_t - \bar{y})^2}} = -0.1014. \tag{3.21}$$

Now the calibration process for the joint modelling of the share prices and the property prices is done. To summarise the results we present the (calibrated) SDEs again:

$$dS(t) = 0.0736 \cdot S(t)dt + 0.1964 \cdot S(t)dW^1(t), \quad S_0 = 1 \tag{3.22}$$

$$dP(t) = 0.0781 \cdot P(t)dt + 0.0444 \cdot P(t)dW^2(t), \quad P_0 = 1 \tag{3.23}$$

with

$$dW^1(t)dW^2(t) = -0.1014 \cdot dt. \tag{3.24}$$

Obviously the expected yearly return on property $\hat{\beta}$ is higher than the corresponding $\hat{\alpha}$ whereas its yearly volatility $\hat{\eta}$ is lower than $\hat{\sigma}$ for the share price process. In this sense property seems to be a more attractive investment than shares. The crucial thing is that property markets are only sparely liquid in practice meaning that our performance index used is maybe a too optimistic proxy. However liquidity squeeze is not considered within this thesis. Figure 3.4 shows a sample simulation for the processes defined in Eqs. 3.22 and 3.23.

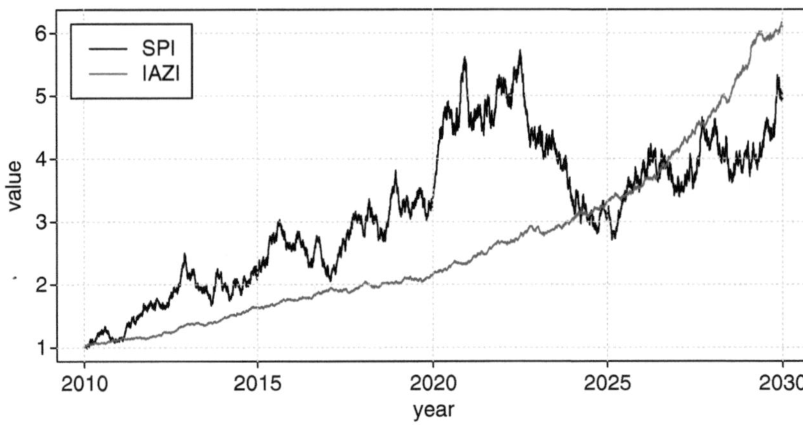

Fig. 3.4 Sample paths for a share index and a property index generated according to two correlated geometric Brownian motions with parameters $\alpha = 0.0736, \sigma = 0.1963, \beta = 0.0781, \eta = 0.0444, \rho = -0.1014$ and $dt = \frac{5}{1{,}257}$

3.3.2 *Generating Stochastic Scenario Files for the Share Price Process and the Property Price Process*

Remember the objective to describe the evolution of all balance sheet entries over time. This issue is handled by Eqs. 3.22 and 3.23. For modelling purposes we will generate a sample of 5,000 simulations for each process and store the year end values for the upcoming 60 years. Thus, that is the scenario environment with pairwise simulation paths for both asset classes needed.

Basically the stochastic scenario file is just a large spreadsheet containing the simulated values for each process. Table 3.4 presents a snapshot of such a file. Note that the prices refer to one unity of the respective index and hence they are initially scaled to 1 as at the end of 2009.

To check the goodness of fit one might recalculate the estimators from the sample and compare these with the input values. Since the sample I have used for the check applies a yearly time convention and not a daily time convention (dt), the resulting differences were pretty satisfying. Confer the results of this check in Table 3.5.

Table 3.4 Snapshot of a stochastic scenario file for the share and the property price process

	Year							
	2009	2010	2011	2012	2013	2014	...	
Simulation 1	SPI	1	0.7429	0.5834	0.5811	0.6274	0.7353	...
	IAZI	1	1.1098	1.1001	1.2211	1.3119	1.3757	...
Simulation 2	SPI	1	0.9256	1.0192	0.8206	0.8721	0.5685	...
	IAZI	1	1.1156	1.1582	1.2018	1.3197	1.4338	...
Simulation 3	SPI	1	1.0287	1.3313	1.5511	1.8650	2.1125	...
	IAZI	1	1.1051	1.2992	1.3804	1.4909	1.4781	...
Simulation 4	SPI	1	0.9968	1.5065	1.8170	2.1448	2.0930	...
	IAZI	1	0.9891	1.0984	1.1873	1.2169	1.4630	...
Simulation 5	SPI	1	0.5989	0.7656	0.7796	0.6855	0.6673	...
	IAZI	1	1.0682	1.1042	1.1842	1.3360	1.3567	...
	...	...	...	...	...	...	...	...

Table 3.5 Comparison of the input parameters with the in-sample estimators of the stochastic scenario files for the share and the property price process

Parameter	Input	Sample	Rel. difference
$\hat{\alpha}$	0.0736	0.0743	−0.009
$\hat{\sigma}$	0.1963	0.1951	0.0062
$\hat{\beta}$	0.0781	0.0780	0.0013
$\hat{\eta}$	0.0444	0.0441	0.0068
$\hat{\rho}$	−0.1014	−0.1008	0.0059

3.4 Basics of Interest Rate Theory

In Sect. 3.3 I have outlined how to model share prices and property prices stochastically. Considering Table 3.3 we see that there is another class to be modelled, *zero coupon bonds* (ZCB) or, to be more precise, ZCB prices. Unlike shares or property, ZCB prices cannot be modelled according to a geometric Brownian motion containing some constant drift variable. Commonly, interest rate modelling is based on the dynamics of some (to be further defined) interest rate r and such are the ZCBs. That rate is typically assumed to be mean reverting and thus it is some kind of opposite of having a constant drift. However the object of this section is to provide the reader with the basics of interest rate theory before defining the requested models to map the bond price process in Sect. 3.5.

The reader can find a general introduction to interest rate theory in Hull (2006). A focus on a deeper mathematical treatment of the theory is provided in Brigo and Mercurio (2006) which I will follow in the subsequent.

3.4.1 *The Short Rate*

Supposing, for example, depositing a certain amount of money in a bank account. Then it is quite natural to expect that amount to grow at some rate as time goes by. The reason for this expectation can be explained by the fact that one unit of money today is considered to be worthier than one unit of money at some time in the future. In this thesis, however, I will not further discuss the legitimateness of that expectation but take it as given. In mathematical terms the idea of the bank account can be formulated as follows.

Definition 7 (bank account). Let $B(t)$ be the value of the *bank account* at time $t \geq 0$ with $B(0) = 1$. Then the bank account is assumed to evolve according the following differential equation

$$dB(t) = r_t B(t)dt, \qquad B(0) = 1, \tag{3.25}$$

where r_t is a positive function of time. As a consequence,

$$B(t) = B(0) \exp\left(\int_0^t r_s ds \right). \tag{3.26}$$

In the above definition r_t is the instantaneous risk free rate at which the bank account accrues. It is referred to as *instantaneous spot rate* or briefly as *short rate*.
Another and basically the most important instrument depending on the short rate is the ZCB.

Definition 8 (zero coupon bond (ZCB)). A $T-$maturity zero coupon bond ($T-$bond) is a contract that guarantees its holder the payment of one unit of money at time T, with no intermediate payments. The contract value at time $t < T$ is denoted by $P(t,T)$ and $P(T,T) = 1 \ \forall\, T$.

To define the price of a ZCB we have to know the nature of the short rate. We may distinguish three different cases[8]:

1. The short rate is a deterministic constant $r > 0$. Then using the definition of the ZCB and the methodology of the bank account we must have

$$\begin{aligned} 1 = P(T,T) &= P(t,T) \exp\left(\int_t^T r\, ds \right) \\ &= P(t,T) \exp(r \cdot (T-t)), \end{aligned} \tag{3.27}$$

   that leads to

$$P(t,T) = \exp(-r \cdot (T-t)), \qquad 0 \leq t \leq T. \tag{3.28}$$

2. The short rate is a time-dependent deterministic function $r_t, t \in \mathbb{R}^+$. Then $P(t,T)$ can be determined as in the first case by

[8]This enumeration is taken from Privault (2008), Sect. 4.1.

$$P(t,T) = \exp\left(-\int_t^T r_s\,ds\right), \qquad 0 \le t \le T. \tag{3.29}$$

3. The short rate is a stochastic process $r_t, t \in \mathbb{R}^+$. Now $P(t,T)$ is given by the following conditional expectation under some martingale measure.

$$P(t,T) = \mathbb{E}\left[\exp\left(-\int_t^T r_s\,ds\right)\middle|\mathcal{F}_t\right], \qquad 0 \le t \le T, \tag{3.30}$$

with $\mathcal{F}_t$ some filtration.

Since there are a few more definitions in order, the short rate shall be supposed to be non stochastic for the moment. Later, for modelling purposes, the short rate is assumed to follow a stochastic process.

3.4.2 Forward Interest Rate and Interest Rate Curves

On the financial markets it is quite common to agree at a present time t on a certain interest $f(t,S,T)$ to be applied to either borrowing or lending a certain amount of money during a future period of time $[T,S]$. That interest rate $f(t,T,S)$ is called *forward interest rate* at time t for the expiration date T and maturity S. To determine an arbitrage free value of the forward interest rate we use simple ZCB terminology. Since

$$\begin{aligned} P(t,S) &= \exp\left(-\int_t^S r_s\,ds\right) = \exp\left(-\int_t^T r_s\,ds\right)\cdot\exp\left(-\int_T^S r_s\,ds\right) \\ &= P(t,T)P(T,S), \end{aligned} \tag{3.31}$$

we have

$$P(T,S) = \frac{P(t,S)}{P(t,T)}, \qquad t \le T \le S. \tag{3.32}$$

The today's prices for a $S-$bond and a $T-$bond are known and so $P(T,S)$ is, assuming $P(t,T) \neq 0$, well-defined. This leads to the definition of the forward interest rate.

Definition 9 (forward interest rate). Let $P(t,T)$ and $P(t,S)$ be the today's price for a ZCB maturing in T and S respectively with $t \le T \le S$. Then the *forward interest rate* $F(t,T,S)$ at time t for the expiration date T and maturity S is defined as

$$F(t,T,S) = -\frac{\log\left(\frac{P(t,S)}{P(t,T)}\right)}{S-T}, \tag{3.33}$$

where log denotes the natural logarithm.

The forward interest rate can also be written as a *spot* (forward interest) rate by $F(t,S) := F(t,t,S)$. Taking the origin definition of the forward interest rate, $F(t,T,S)$ again and letting $S \searrow T$, the forward interest rates converges to the *instantaneous forward interest rate*.

Definition 10 (instantaneous forward interest rate). Let $P(t,T)$ be the today's price for a ZCB maturing in T with $t \leq T$. Then the *instantaneous forward interest rate* $f(t,T)$ at time t for the maturity date T is defined as

$$f(t,T) := \lim_{S \searrow T} F(t,T,S) = -\frac{\partial \log P(t,T)}{\partial T}. \tag{3.34}$$

Sometimes it is useful to consider for a fixed point in time t ZCB prices as a function of their maturity T. Hence given the time t price of any ZCB, $P(t,T)$, with maturity T we can define by

$$T \mapsto P(t,T), \qquad T \geq t, \tag{3.35}$$

the *zero bond curve* at time t. Such a curve is also referred to as a *term structure of discount factors*. Assuming that interest rates are strictly positive, the zero bond curve is decreasing in T. Figure 3.5 shows the zero bond curve for *Swiss confederation bonds* as at the end of 2009. Note that generally bond prices are

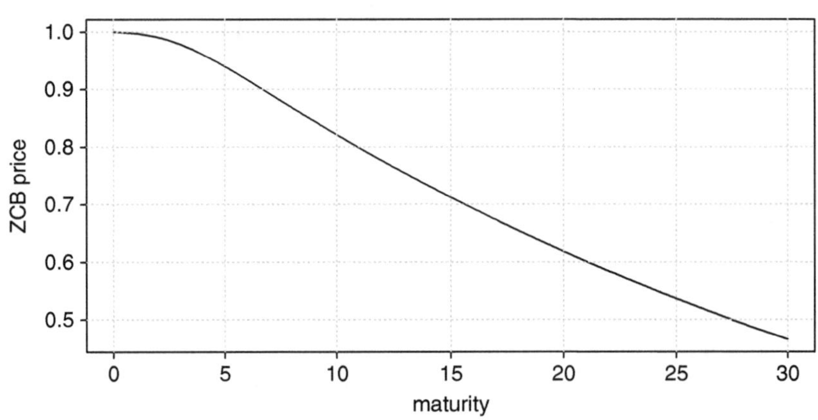

Fig. 3.5 Term structure of discount factors stripped from Swiss confederation bonds as at the end of 2009 (Source: Datastream)

observable for discrete time points only. So to smoothen the curve in that figure a cubic spline interpolation was applied performing `spline(...)` in **R**.[9]

The second curve to be introduced is the *instantaneous forward rate curve* given by

$$T \mapsto f(t,T), \qquad T \geq t. \tag{3.36}$$

See Fig. 3.6 showing the instantaneous forward rate curve stripped from Swiss confederation bonds as at the end of 2009. Again, the curve is smoothened by a cubic spline interpolation.

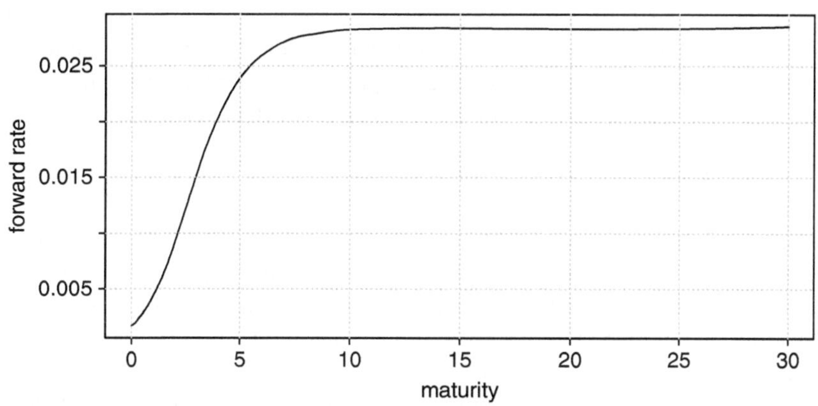

Fig. 3.6 Instantaneous forward rate curve stripped from Swiss confederation bonds as at the end of 2009 (Source: Datastream)

After this brief introduction to the basics of interest rate theory the reader is now well prepared for the modelling of the bond price process. In the upcoming section I will introduce three different interest rate models. Each one has, in theory, its advantages. It will be a future aspect then to see what various impacts these models have in practice when doing the empirical analysis in Chap. 5.

3.5 The Bond Price Process

Recall the proxy balance sheet in Table 3.3. More than 78% of the total assets amount are allocated to the class of ZCB prices. Thus, we definitely have to deal with the modelling of that class with great care. Basically I will proceed in the same manner as for the first two classes, shares and property. Once the stochastic process has been defined it needs to be calibrated to observable market data. Having done

[9]For further information see The R Development Core Team (2010).

this, a set of 5,000 stochastic ZCB prices evolving over 50 years will be generated. Note that here one simulation consists of multiple ZCB prices of different maturities in order to match, for example, the structure of a given bond portfolio.

During the last two decades the theory of interest rate modelling has grown enormously and so selecting the most appropriate one for our purposes from the vast of feasible models becomes a challenging issue. A priori it is not obvious which one best fits our needs, so we heuristically choose three different approaches, a single factor Cox-Ingersoll-Ross (CIR) model, a single factor Hull-White (HW) model and a two-additive-factor Gaussian (G2++) model.

In the subsequent I will follow Brigo and Mercurio (2006) to introduce each of these interest rate models followed by its calibration. Afterwards a summarising comparison of the calibration results is presented in Sect. 3.5.4. In 1977 O. A. Vasicek presented his equilibrium interest rate model.[10] To become familiar with interest rate models in general this is a good paper to start from. There the dynamics of the short rate r is assumed to follow an Ornstein-Uhlenbeck process

$$dr(t) = k[\theta - r(t)]dt + \sigma dW(t), \qquad r(0) = r_0, \tag{3.37}$$

under the risk-neutral measure with r_0, k, θ and $\sigma \in \mathbb{R}^+$ and W a Brownian motion.

The Vasicek model may be accounted as the pioneering work of the class of endogenous term structure models where the current term structure of interest rates is rather an output than an input. Even though that model is very tractable it has at least two notable drawbacks: Firstly, it is not able to fit the current term structure satisfactorily and secondly, the short rate r can become negative with a positive probability. An extension of the Vasicek model that allows for non-negative interest rates only was proposed by Cox, Ingersoll and Ross in 1985.[11]

3.5.1 The Cox-Ingersoll-Ross (CIR) Model

The risk-neutral process of the short rate r in the CIR model is given by the following SDE:

$$dr(t) = k[\theta - r(t)]dt + \sigma\sqrt{r(t)}dW(t), \quad r(0) = r_0, \tag{3.38}$$

with r_0, k, θ and $\sigma \in \mathbb{R}^+$ and W a Brownian motion. Requesting for

$$2k\theta > \sigma^2 \tag{3.39}$$

ensures that $r(t)$ remains positive for all t.

The mean and the variance of $r(t)$ conditional on the filtration $\mathcal{F}_s$ are given by

[10] Confer Vašiček (1977).

[11] See Cox et al. (1985).

$$\mathbb{E}[r(t)|\mathcal{F}_s] = r(s)e^{-k(t-s)} + \theta\left(1 - e^{-k(t-s)}\right), \tag{3.40}$$

$$\begin{aligned}\text{Var}(r(t)|\mathcal{F}_s) = r(s)\frac{\sigma^2}{k}\left(e^{-k(t-s)} - e^{-2k(t-s)}\right) \\ +\theta\frac{\sigma^2}{2k}\left(1 - e^{-k(t-s)}\right)^2\end{aligned} \tag{3.41}$$

One can show that in the CIR model the time t price of a ZCB with maturity T, $P(t,T)$, is given by the following formulae.

$$P(t,T) = A(t,T)e^{-B(t,T)r(t)}, \tag{3.42}$$

where

$$\begin{aligned}A(t,T) &= \left[\frac{2h\exp\{(k+h)(T-t)/2\}}{2h+(k+h)(\exp\{(T-t)h\}-1)}\right]^{2k\theta/\sigma^2}, \\ B(t,T) &= \frac{2(\exp\{(T-t)h\}-1)}{2h+(k+h)(\exp\{(T-t)h\}-1)}, \\ h &= \sqrt{k^2+2\sigma^2}.\end{aligned}$$

To calibrate the CIR model to observable market data I propose to minimise the following mean squared error term Ψ.

$$\Psi = \frac{1}{n}\sum_{T=T_1}^{T_n}\left(\frac{P^{\text{CIR}}(0,T)}{P^{\text{market}}(0,T)} - 1\right)^2, \tag{3.43}$$

where $P^{\text{CIR}}(0,T)$ and $P^{\text{market}}(0,T)$ stand for the today's CIR model price of a ZCB maturing in T as defined in Eq. 3.42 and the today's observed market price of a ZCB maturing in T respectively. Indeed we have $P(t,T) = P(t,T,k,\sigma,\theta)$ in the CIR model meaning that we need to minimise across three parameters k,σ and θ, whereas $t = 0$ and T are given from the available data.

As in the previous section, the model is calibrated as at the end of 2009. Regarding potential estimation risks it could be dangerous just to take market prices only from a single day in this case. Instead I will consider month end data for past 12 months. The data basis is given by historical swap rates provided by Datastream. Therefrom ZCB market prices $P^{\text{market}}(0,T_i)$ are derived by

$$P^{\text{market}}(0,T_i) = \exp(-T_i \cdot s_{T_i}), \tag{3.44}$$

with s_{T_i} the today's T_i-year swap rate for $T_i = \{1,2,\ldots,10,12,15,20\}$. In Table A.2 in Appendix A.1 the complete data basis is shown.

I proceed by minimising Ψ with a Nelder-Mead routine that is available in **R**. Let `Psi_CIR.R` be the function to calculate the expression in Eq. 3.43 based on the data as outlined above. The Ψ is minimised in **R** as follows.

```
> params_start <- c( k = 0.5,sigma = 0.1, theta = 0.05)
> optim(params_start, Psi_CIR.R, method="Nelder-Mead")
$par
[1] 0.21498096 0.13289474 0.04108091
$value
[1] 0.0001885126
```

In the first line the staring values for the parameters in the optimisation algorithm with $k = 0.5, \sigma = 0.1$ and $\theta = 0.05$ are defined. These values were set more or less arbitrarily but still in consideration of Constraint 3.39.

With `optim(...)` we finally get the best fit parameter set $k = 0.2150, \sigma = 0.1329$ and $\theta = 0.0411$ that yields a mean squared error $\Psi = 0.000189$. Table 3.6 below shows the calibration results in detail.

Despite the feature that the CIR model only allows for nonnegative values for the short rate r, it still has a major drawback: It does not satisfactorily fit the initial term structure, as this is symptomatic for endogenous term structure models. Referring to this an extension was made when proposing exogenous term structure models. In 1990 Hull and White (1990) improved the Vasicek model by introducing a time-varying parameter to exactly fit the currently observed term structure of interest

Table 3.6 Calibration results for the CIR model as at the end of 2009 in % given by the squared relative difference $(P^{\mathrm{CIR}}(0, T_i)/P^{\mathrm{market}}(0, T_i) - 1)^2$ (Source: Datastream)

Date of observation	Maturity T_i of ZCB in years												
	1	2	3	4	5	6	7	8	9	10	12	15	20
30.12.09	0.001	0.004	0.006	0.007	0.006	0.004	0.002	0.001	0.000	0.000	0.002	0.008	0.001
30.11.09	0.001	0.007	0.010	0.012	0.013	0.012	0.011	0.009	0.008	0.007	0.004	0.002	0.027
30.10.09	0.001	0.004	0.006	0.006	0.004	0.003	0.001	0.001	0.000	0.000	0.004	0.017	0.004
30.09.09	0.001	0.006	0.010	0.011	0.010	0.007	0.005	0.002	0.001	0.000	0.001	0.011	0.004
30.08.09	0.001	0.008	0.014	0.016	0.014	0.011	0.006	0.003	0.001	0.000	0.001	0.011	0.002
30.07.09	0.001	0.007	0.009	0.008	0.005	0.002	0.000	0.001	0.003	0.005	0.016	0.042	0.027
30.06.09	0.001	0.006	0.006	0.002	0.000	0.004	0.015	0.028	0.044	0.057	0.096	0.165	0.193
30.05.09	0.001	0.007	0.010	0.004	0.000	0.005	0.017	0.037	0.068	0.099	0.157	0.238	0.175
30.04.09	0.001	0.006	0.008	0.007	0.004	0.002	0.001	0.000	0.000	0.000	0.000	0.004	0.139
30.03.09	0.001	0.007	0.011	0.009	0.005	0.002	0.000	0.000	0.000	0.001	0.002	0.000	0.177
28.02.09	0.001	0.006	0.008	0.007	0.003	0.001	0.000	0.000	0.000	0.000	0.001	0.002	0.231
30.01.09	0.001	0.007	0.010	0.011	0.009	0.007	0.006	0.005	0.004	0.004	0.005	0.019	0.276

rates. I will introduce their model by presenting the special case as analysed in Hull and White (1994).

3.5.2 The Hull-White (HW) Model

The Hull-White (HW) model assumes that the short rate process evolves under the risk neutral measure according to

$$dr(t) = [\vartheta(t) - ar(t)]dt + \sigma dW(t), \tag{3.45}$$

where a and $\sigma \in \mathbb{R}^+, \vartheta(t)$ a deterministic function of time and W a Brownian motion. To achieve an exact fit of the current term structure, $\vartheta(t)$ must be chosen appropriately. Let $f^M(0,t)$ denote the today's instantaneous forward interest rate for the maturity date t. The subscript M indicates that f^M is a (observable) market interest rate. Then

$$\vartheta(t) = \frac{\partial f^M(0,t)}{\partial t} + af^M(0,t) + \frac{\sigma^2}{2a}(1 - e^{-2at}) \tag{3.46}$$

ensures the desired fit of the current term structure.

The HW model provides a short rate r that is conditional on the filtration $\mathcal{F}_s$ normally distributed with mean and variance given respectively by

$$\mathbb{E}[r(t)|\mathcal{F}_s] = r(s)e^{-a(t-s)} + \alpha(t) - \alpha(s)e^{-a(t-s)}, \tag{3.47}$$

$$\text{Var}(r(t)|\mathcal{F}_s) = \frac{\sigma^2}{2a}\left[1 - e^{-2a(t-s)}\right], \tag{3.48}$$

where

$$\alpha(t) = f^M(0,t) + \frac{\sigma^2}{2a^2}(1 - e^{-at})^2.$$

A clear drawback of the current model is the positive probability of negative values for the short rate. Using the stochastic process defined in Eq. 3.45 the risk neutral probability of negative interest rates at time t can explicitly be calculated by

$$P(r(t) < 0) = \Phi\left(-\frac{\alpha(t)}{\sqrt{\frac{\sigma^2}{2a}[1 - e^{-2at}]}}\right) \tag{3.49}$$

with Φ denoting the standard normal cumulative distribution function. However Brigo and Mercurio (2006) comment that probability as *almost negligible in practice*.

In the HW model the price of a ZCB at time t with maturity T, P(t,T), is given by

$$P(t,T) = A(t,T)e^{-B(t,T)r(t)}, \tag{3.50}$$

where

$$B(t,T) = \frac{1}{a}\left[1 - e^{-a(T-t)}\right],$$

$$A(t,T) = \frac{P^M(0,T)}{P^M(0,t)} \exp\left\{B(t,T)f^M(0,t) - \frac{\sigma^2}{4a}(1-e^{-2at})B(t,T)^2\right\},$$

and P^M denoting the market ZCB price.

As suggested in Brigo and Mercurio (2006) I will calibrate the HW model to a set of *at the money* (ATM) payer swaption prices. For our purposes we represent the price of a payer swaption as a sum of put options on a ZCB.

In terms of the HW model the price of a European put option $ZBP(t,T,S,X)$ at time t with maturity T and strike price X on a ZCB maturing in S is given by

$$ZBP(t,T,S,X) = XP(t,T)\Phi(-h+\sigma_p) - P(t,S)\Phi(-h), \tag{3.51}$$

where

$$\sigma_p = \sigma\sqrt{\frac{1-e^{-2a(T-t)}}{2a}}B(T,S),$$

$$h = \frac{1}{\sigma_p}\log\frac{P(t,S)}{P(t,T)X} + \frac{\sigma_p}{2},$$

and P, Φ, a and B as defined above. Given that formula for the put option on a ZCB we can also determine the price of a European payer swaption.

Consider a European payer swaption with strike rate X, maturity T and nominal value N, which gives the holder the right to enter at time $t_0 = T$ an interest rate swap with payment times $\mathcal{T} = \{t_1, \ldots, t_n\}$, $t_1 > T$, where he pays at the fixed rate X and receives LIBOR set in arrears. τ_i denotes the year fraction from t_{i-1} to t_i, $i = 1, \ldots, n$ and set $c_i = X\tau_i$ for $i = 1, \ldots, n-1$ and $c_n = 1 + X\tau_n$. Denoting by r^* the value of the short rate at time T for which

$$\sum_{i=1}^{n} c_i A(T,t_i)e^{-B(T,t_i)r^*} = 1,$$

and setting $X_i = A(T, t_i) \exp -B(T, t_i) r^*$, the price of the payer swaption at time $t < T$, $PS(t, T, \mathcal{T}, N, X)$, is then given by

$$PS(t, T, \mathcal{T}, N, X) = N \sum_{i=1}^{n} c_i ZBP(t, T, t_i, X_i). \tag{3.52}$$

So with Eq. 3.52 we have the formula to calculate the price of a payer swaption according to the HW model. To finally formulate a constraint in order to calibrate the HW model we still need the respective market prices.

Datastream does not directly provide swaption prices but swaption implied volatilities $\sigma_{m,s}$ for a given maturity m and s the length of the underlying swap. The complete data basis can be found in Table A.1 in Appendix A.1. Taking one of Black's formulas we can derive today's prices of any payer swaption as if it were market prices.[12]

In order to generate swaption market prices we define for each given swaption implied volatilitiy $\sigma_{m,s}$ the following setting. We assume the tenor structure to be equal to $\mathcal{T} = \{m + 0.5, m + 1, \ldots, m + s\}$ with half-yearly settlements meaning that $\tau_i = 0.5$ for all i. Furthermore there is K the strike price of the option, $S_{m,s}(0)$ the current forward swap rate for an $s-$year swap starting in m years and $N = 1$ the nominal value. Then the today's ATM price of a payer swaption with maturity m on a s year swap is given by

$$\begin{aligned} PS^{\text{Black}}(0, \mathcal{T}, N, K, \sigma_{m,s}) &= \\ & N\text{Bl}(K, S_{m,s}(0), \sigma_{m,s}\sqrt{m}, 1) \sum_{i \in \mathcal{T}} 0.5 \cdot P(0, i) \end{aligned} \tag{3.53}$$

where Bl(...) denotes Blacks option pricing formula. Setting

$$K = K_{\text{ATM}} = S_{m,s}(0) = \frac{P(0, m) - P(0, m + s)}{\sum_{i \in \mathcal{T}} 0.5 \cdot P(0, i)} \tag{3.54}$$

the price calculated according to Formula 3.53 becomes at the market.

As shown in Table A.1 in Appendix A.1 we have a set of at the money European implied swaption volatility quotes (CHF) $\sigma_{m,s}$ as at the end of 2009 for option maturities $m \in M = \{1, 2, 3, 4, 5, 7, 10\}$ years on swaps of length $s \in S = \{1, 2, \ldots, 10\}$ years. Using the above introduced Formulae 3.53 and 3.54 the swaption implied volatilities are taken to generate our today's observed market prices. Then the target function of the calibration procedure is as follows.

Let $PS^{\text{market}}(m, s) = PS^{\text{Black}}(0, \mathcal{T}, 1, S_{m,s}(0), \sigma_{m,s})$ be the today's observed market price of an at the money payer swaption with maturity m on a $s-$year swap. The swap is assumed to be settled twice a year with a strike rate $S_{m,s}(0)$,

[12] Confer Brigo and Mercurio (2006), pp. 19–22.

the notional $N = 1$ and its implied volatility shall be $\sigma_{m,s}$. Analogously we denote the respective HW model price by $PS^{\text{HW}}(m,s) = PS(0,m,\mathcal{T},1,S_{m,s}(0),a,\sigma)$ as defined in Eq. 3.50. Then the target function to be minimised is again given by a mean squared error term Ψ.

$$\Psi = \frac{1}{|S| \cdot |M|} \sum_{m \in M, s \in S} \left(\frac{PS^{\text{HW}}(m,s)}{PS^{\text{market}}(m,s)} - 1 \right)^2 \tag{3.55}$$

Note that calibrating the HW model to current market payer swaption prices means determining the parameters α and σ appropriately in the function $PS(0,m,\mathcal{T},1,S_{m,s}(0),a,\sigma)$ such that Ψ gets minimised. Since m and s are given, the remaining quantities $\mathcal{T}$ and $S_{m,s}(0)$ are implicitly defined by the assumptions made on the terms of the market payer swaptions.

The calibration is done in **R** again by applying the routine `optim(...)` on `Psi_HW.R` that minimises the term in Eq. 3.55 using a Nelder-Mead optimising algorithm. The data basis used was outlined above.

```
> params_start <- c(a = 0.1, sigma = 0.01)
> optim(params_start, Psi_HW, method="Nelder-Mead")

$par
[1] 0.01671814 0.00789167
$value
[1] 0.00336253
```

The routine finds the optimal parameters $a = 0.0167$ and $\sigma = 0.0079$ that yield a $\Psi = 0.00336$.

Figure 3.7 gives a detailed overview on the calibration results for the HW model by the squared relative difference $(PS^{\text{HW}}(m,s)/PS^{\text{market}}(m,s) - 1)^2$.

In the following section I will introduce the last interest rate model to be considered: The Two Additive Factor Gaussian Model.[13] The main advantage of a two factor or a multi factor model in general is, in this case, its ability to capture more sophisticated volatility structures and correlation patterns. Thus, one might expect a better calibration fit.

On the other hand, the implementation of a multi factor model often implies a substantial amount of computational work and the models further need, by trend, a larger calibration data basis in order to avoid an overfitting. One can interpret this as the drawbacks of a multi factor model.

[13] See Brigo and Mercurio (2006).

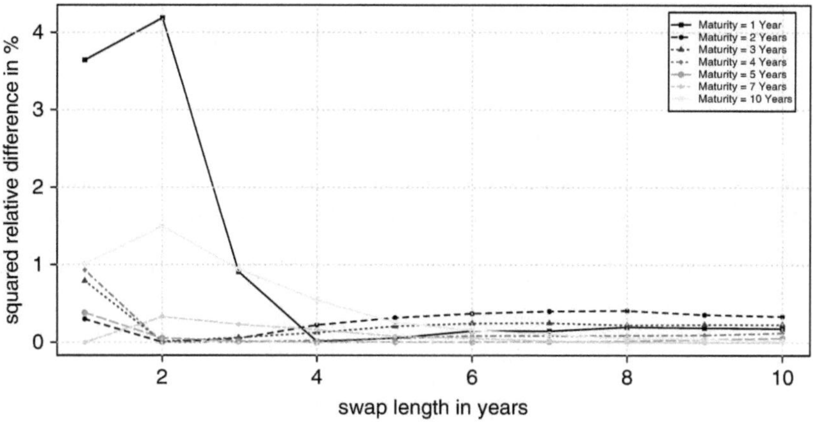

Fig. 3.7 Calibration results for the HW model as at the end of 2009 in % given by the squared relative difference $(PS^{\mathrm{HW}}(m,s)/PS^{\mathrm{market}}(m,s)-1)^2$

3.5.3 *The Two Additive Factor Gaussian (G2++) Model*

The process for the short rate in the two additive factor Gaussian (G2++) model is given by the sum of two correlated Gaussian factors plus a deterministic function. If the latter one is chosen appropriately the model exactly fits the current term structure of interest rates.

Let us have a look at the mathematics behind the model. The process for the short rate r under the risk-adjusted measure Q is given by the following equations.

$$r(t) = x(t) + y(t) + \varphi(t), \qquad r(0) = r_0, \tag{3.56}$$

where the processes $\{x(t) : t \geq 0\}$ and $\{y(t) : t \geq 0\}$ satisfy

$$dx(t) = -ax(t)dt + \sigma dW^1(t), \qquad x(0) = 0, \tag{3.57}$$

$$dy(t) = -by(t)dt + \eta dW^2(t), \qquad y(0) = 0, \tag{3.58}$$

where (W^1, W^2) is a two-dimensional Brownian motion with instantaneous correlation ρ as from

$$dW^1(t)dW^2(t) = \rho dt, \tag{3.59}$$

where r_0, a, b, σ and η are positive constants and $-1 \leq \rho \leq 1$. As mentioned above there is a deterministic function, φ, enabling an exact fit of the current term structure if chosen appropriately. We achieve so with

$$\begin{aligned}\varphi(t) &= f^M(0,t) + \frac{\sigma^2}{2a^2}(1-e^{-at})^2 + \frac{\eta^2}{2b^2}(1-e^{-bt})^2 \\ &+\rho\frac{\sigma\eta}{ab}(1-e^{-at})(1-e^{-bt}).\end{aligned} \tag{3.60}$$

Under these circumstances the short rate r in the G2++ model is conditional on the filtration $\mathcal{F}_s$ normally distributed with mean and variance given respectively by

$$\mathbb{E}[r(t)|\mathcal{F}_s] = x(s)e^{-a(t-s)} + y(s)e^{-b(t-s)} + \varphi(t), \tag{3.61}$$

$$\begin{aligned}\mathrm{Var}(r(t)|\mathcal{F}_s) &= \frac{\sigma^2}{2a}\left[1-e^{-2a(t-s)}\right] + \frac{\eta^2}{2b}\left[1-e^{-2b(t-s)}\right] \\ &+2\rho\frac{\sigma\eta}{a+b}\left[1-e^{-(a+b)(t-s)}\right].\end{aligned} \tag{3.62}$$

Applying Eq. 3.60 implies that the risk-neutral probability of negative rates for r at time t is

$$P(r(t)<0) = \Phi\left(-\frac{\mathbb{E}[r(t)]}{\sqrt{\mathrm{Var}(r(t))}}\right), \tag{3.63}$$

which is by the words of D. Brigo and F. Mercurio *often negligible in many concrete situations.*

Within the G2++ model ZCBs can be priced with a closed form solution. The time t price for a ZCB with maturity T, $P(t,T)$, is calculated according to

$$P(t,T) = \frac{P^M(0,T)}{P^M(0,t)}\exp\{\mathcal{A}(t,T)\}, \tag{3.64}$$

where

$$\begin{aligned}\mathcal{A}(t,T) &= \frac{1}{2}[V(t,T) - V(0,T) + V(0,t)] \\ &-\frac{1-e^{-a(T-t)}}{a}x(t) - \frac{1-e^{-b(T-t)}}{b}y(t), \\ V(t,T) &= \frac{\sigma^2}{a^2}\left[T-t+\frac{2}{a}e^{-a(T-t)} - \frac{1}{2a}e^{-2a(T-t)} - \frac{3}{2a}\right] \\ &+\frac{\eta^2}{b^2}\left[T-t+\frac{2}{b}e^{-b(T-t)} - \frac{1}{2b}e^{-2b(T-t)} - \frac{3}{2b}\right] \\ &+2\rho\frac{\sigma\eta}{ab}\left[T-t+\frac{e^{-a(T-t)}-1}{a} + \frac{e^{-b(T-t)}-1}{b}\right. \\ &\left.-\frac{e^{-(a+b)(T-t)}-1}{a+b}\right].\end{aligned}$$

To calibrate the G2++ model I will proceed in a similar manner to the one factor HW model. On the one hand market prices for European payer swaptions are needed. The data basis for this shall be a set of at the money European implied swaption volatility quotes (CHF) as at the end of 2009 again, whereof market prices of European payer swaptions are derived.[14] A detailed description of this procedure was outlined in Sect. 3.5.2. On the other hand, we have to calculate equivalent model prices for the G2++ model. Here a closed form solution no longer exists but instead, a numerical solution has to be applied.

Consider a European payer swaption with strike rate X, maturity T and nominal value N, which gives the holder the right to enter at time $t_0 = T$ an interest rate swap with payment times $\mathcal{T} = \{t_1, \dots, t_n\}$, $t_1 > T$, where he pays at the fixed rate X and receives LIBOR set in arrears. τ_i denotes the year fraction from t_{i-1} to t_i, $i = 1, \dots, n$ and set $c_i = X\tau_i$ for $i = 1, \dots, n-1$ and $c_n = 1 + X\tau_n$. The price at time $t = 0$ of that payer swaption $PS(0, T, \mathcal{T}, N, X)$ is then given by numerically computing of the following one-dimensional integral.

$$
\begin{aligned}
PS(0,T,\mathcal{T},N,X) &= \qquad (3.65) \\
NP(0,T) \int_{-\infty}^{\infty} & \frac{e^{-\frac{1}{2}\left(\frac{x-\mu_x}{\sigma_x}\right)^2}}{\sigma_x\sqrt{2\pi}} \left[\Phi(-h_1(x)) - \sum_{i=1}^{n} \lambda_i(x) e^{\kappa_i(x)} \Phi(-h_2(x)) \right] dx,
\end{aligned}
$$

with

$$
\begin{aligned}
h_1(x) &= \frac{\bar{y}-\mu_y}{\sigma_y\sqrt{1-\rho_{xy}^2}} - \frac{\rho_{xy}(x-\mu_x)}{\sigma_x\sqrt{1-\rho_{xy}^2}}, \\
h_2(x) &= h_1(x) + B(b,T,t_i)\sigma_y\sqrt{1-\rho_{xy}^2}, \\
\lambda_i(x) &= c_i A(T,t_i) e^{-B(a,T,t_i)x}, \\
\kappa_i(x) &= -B(b,T,t_i)\left[\mu_y - \frac{1}{2}(1-\rho_{xy}^2)\sigma_y^2 B(b,T,t_i) + \rho_{xy}\sigma_y \frac{x-\mu_x}{\sigma_x} \right], \\
A(t,T) &= \frac{P^M(0,T)}{P^M(0,t)} \exp\left\{ \frac{1}{2}[V(t,T) - V(0,T) + V(0,t)] \right\}, \\
B(z,T,S) &= \frac{1-e^{-z(S-T)}}{z}.
\end{aligned}
$$

$\bar{y} = \bar{y}(x)$ is the unique solution of the following equation.

[14] See Table A.1 in Appendix A.1.

$$\sum_{i=1}^{n} c_i A(T,t_i) e^{-B(a,T,t_i)x - B(b,T,t_i)\bar{y}} = 1 \tag{3.66}$$

and

$$\begin{aligned}
\mu_x &= -M_x^T(0,T),\\
\mu_y &= -M_y^T(0,T),\\
\sigma_x &= \sigma\sqrt{\frac{1-e^{-2aT}}{2a}},\\
\sigma_y &= \eta\sqrt{\frac{1-e^{-2bT}}{2b}},\\
\rho_{xy} &= \frac{\rho\sigma\eta}{(a+b)\sigma_x\sigma_y}\left[1-e^{-(a+b)}T\right],
\end{aligned}$$

where

$$\begin{aligned}
M_x^T(s,t) &= \left(\frac{\sigma^2}{a^2}+\rho\frac{\sigma\eta}{ab}\right)\left[1-e^{-a(t-s)}\right]-\frac{\sigma^2}{2a^2}\left[e^{-a(T-t)}-e^{-a(T+t-2s)}\right]\\
&\quad -\frac{\rho\sigma\eta}{b(a+b)}\left[e^{-b(T-t)}-e^{-bT-at+(a+b)s}\right],\\
M_y^T(s,t) &= \left(\frac{\eta^2}{b^2}+\rho\frac{\sigma\eta}{ab}\right)\left[1-e^{-b(t-s)}\right]-\frac{\eta^2}{2b^2}\left[e^{-b(T-t)}-e^{-b(T+t-2s)}\right]\\
&\quad -\frac{\rho\sigma\eta}{a(a+b)}\left[e^{-a(T-t)}-e^{-aT-bt+(a+b)s}\right].
\end{aligned}$$

Now we have defined both, the market price and the G2++ model price of a European payer swaption, PS^{market} and $PS^{\text{G2++}}$ respectively. To calibrate the G2++ model to European payer swaption prices minimise the following expression.

$$\Psi = \frac{1}{|S|\cdot|M|}\sum_{m\in M, s\in S}\left(\frac{PS^{\text{G2++}}(m,s)}{PS^{\text{market}}(m,s)}-1\right)^2, \tag{3.67}$$

with $PS^{\text{G2++}}(m,s) = PS(0,m,\mathcal{T},1,S_{m,s}(0),a,b,\sigma,\eta,\rho)$ as defined in Eq. 3.65 and S, M and PS^{market} as defined in Sect. 3.5.2. The strike rate $S_{m,s}(0)$ of both swaptions is chosen such that the prices become ATM.

So there is a lot of computational work behind the formulae to price a swaption according the G2++ model. This already causes a huge amount of runtime. Additionally, by a, b, σ, η and ρ there are three more parameters to be determined compared to the CIR model or the HW model. Keeping in mind these arguments and knowing that the model responds highly sensitively to the starting parameters, one realises that the calibration procedure definitely becomes a delicate issue here. To avoid getting lost we make use of an approximative algorithm to price swaptions presented by Schrager and Pelsser (2006) that is significantly faster than the origin G2++ algorithm. The idea behind that algorithm is to find values for a, b, σ, η and ρ by applying the so called Schrager algorithm that can be further used as reasonable starting values in the G2++ algorithm.

In the subsequent I will briefly introduce the Schrager algorithm that was originally presented in Schrager and Pelsser (2006). The formulae stated below are an alternative solution as derived in van Haastrecht (2007).

Suppose a European payer swaption with strike rate $S_{\alpha,\beta}(0)$ such that the swaption becomes ATM. Its maturity shall be T_α, its nominal $N = 1$. The swaption gives the holder the right to enter at $t = T_{\alpha+1}$ a swap with payment times $\mathcal{T} = \{T_{\alpha+1}, \ldots, T_{\alpha+n=\beta}\}$ where he pays at the fixed rate $S_{\alpha,\beta}(0)$ and receives LIBOR set in arrears. τ_i denotes the year fraction from $T_{\alpha+i-1}$ to $T_{\alpha+i}$, $i = 1, \ldots, n$. Then the price of the swaption $PS^{\text{Schrager}}(0, T_\alpha, \mathcal{T}, 1, S_{\alpha,\beta}(0))$ can be calculated according to the following formulae.

$$PS^{\text{Schrager}}(0, T_\alpha, \mathcal{T}, N = 1, S_{\alpha,\beta}(0)) = \mathcal{N}_{\alpha+1,\beta}(0)\frac{\sigma_{\alpha,\beta}}{\sqrt{2\pi}}, \tag{3.68}$$

where

$$\mathcal{N}_{\alpha+1,\beta}(0) = \sum_{i=\alpha+1}^{\beta} \tau_i P(0, T_i),$$

$$\sigma_{\alpha,\beta} = \Bigg(\sigma^2 (C^{(1)}_{\alpha,\beta})^2 \left[\frac{e^{2aT_\alpha} - 1}{2a}\right] + \eta^2 (C^{(2)}_{\alpha,\beta})^2 \left[\frac{e^{2bT_\alpha} - 1}{2b}\right] + 2\rho\sigma\eta C^{(1)}_{\alpha,\beta} C^{(2)}_{\alpha,\beta} \left[\frac{e^{(a+b)T_\alpha} - 1}{a+b}\right]\Bigg)^{\frac{1}{2}},$$

$$C^{(1)}_{\alpha,\beta} = \frac{1}{a}\Bigg[e^{-aT_\alpha} P^{\mathcal{N}}(0, T_\alpha) - e^{-aT_\beta} P^{\mathcal{N}}(0, T_\beta) - S_{\alpha,\beta}(0) \sum_{i=\alpha+1}^{\beta} \tau_i e^{-aT_i} P^{\mathcal{N}}(0, T_i)\Bigg],$$

$$C^{(2)}_{\alpha,\beta} = \frac{1}{b}\Big[e^{-bT_\alpha} P^{\mathcal{N}}(0,T_\alpha) - e^{-bT_\beta} P^{\mathcal{N}}(0,T_\beta)$$

$$-S_{\alpha,\beta}(0) \sum_{i=\alpha+1}^{\beta} \tau_i e^{-bT_i} P^{\mathcal{N}}(0,T_i)\Big],$$

$$P^{\mathcal{N}}(t,T_i) = \frac{P(t,T_i)}{\mathcal{N}_{\alpha+1,\beta}(t)}.$$

Recall the data basis to calibrate the model. There we have a set of swaption implied volatilities refering to a swaptions maturing in $m \in M = \{1,2,3,4,5,7,10\}$ years on a swap of length $s \in S = \{1,2,\ldots,10\}$ years. Thus for each set of (m,s) we can calculate an approximative model price for the payer swaption according to Formula 3.68 by setting $T_\alpha = m$ and $\mathcal{T} = \{T_{\alpha+1} = m + 0.5, T_{\alpha+2} = m + 1,\ldots,T_\beta = m + s\}$. Note that $\tau_i = 0.5$ and still $N = 1$. Now suppose a **R** routine `Psi_G2_approx.R` that calculates Ψ in Eq. 3.67 assuming the model value $PS^{\text{G2++}}$ being replaced by the approximative value PS^{Schrager} given by the Schrager algorithm.

```
> params_start <- c(a = 0.8,b = 0.2,sigma = 0.01,
                    eta = 0.01,rho = -0.8)
> optim(params_start, Psi_G2_approx,
        method="Nelder-Mead")

$par
[1] 1.0949141 0.0395176 0.0063241 0.0092879, -0.9989981
$value
[1] 0.002270143
```

The solution of this minimisation yields a temporary best fit parameter set $(a = 1.0949, b = 0.0395, \sigma = 0.0063, \eta = 0.0093, \rho - 0.9990)$ that shall be the starting values for the final calibration with the origin G2++ pricing formula. Before doing that last step we shall have a look at the quality of the approximative prices. Given the temporary best fit parameter set, we are going to consider the ratios $(PS^{\text{Schrager}}(m,s)/PS^{\text{G2++}}(m,s) - 1)$. See Fig. 3.8 therefor.

Given a mean value for the relative differences of 0.04%, we constitute that the Schrager algorithm indeed provides an acceptable approximation of the G2++ swaption prices. The final calibration is then started from this temporary best fit parameter set determined by the Schrager algorithm. The function to be optimised is coded in `Psi_G2.R` that calculates the Ψ according to Eq. 3.67.

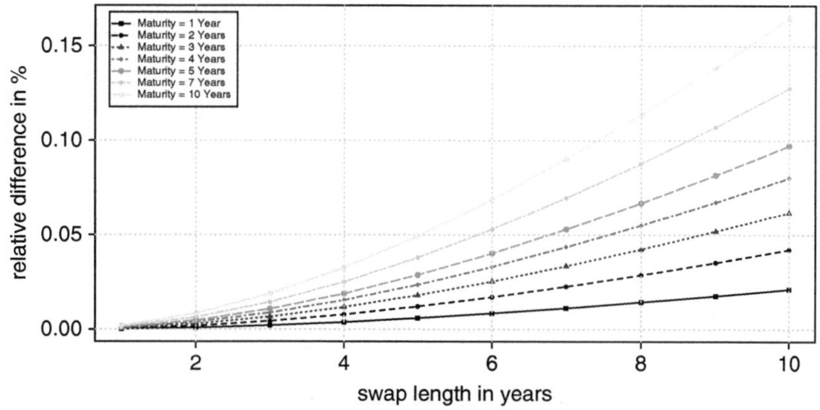

Fig. 3.8 The quality of the Schrager algorithm measured by the ratios $(PS^{\text{Schrager}}(m,s)/PS^{\text{G2++}}(m,s) - 1)$ in % based on the temporary best fit parameters $a = 1.0949, b = 0.0395, \sigma = 0.0063, \eta = 0.0093$ and $\rho - 0.9990$

```
> params_start <- c(1.0949141, 0.0395176, 0.0063241,
                    0.0092879, -0.9999981)
> optim(params_start, Psi_G2,
        method="Nelder-Mead")
> $par
> [1]2.366030868 0.032710346 0.016622460
     0.008876422 -0.946196095
> $value
> [1] 0.002248167
```

This time we get the final parameters for the G2++ model by $a = 2.3660, b = 0.0327, \sigma = 0.0166, \eta = 0.0089$ and $\rho = -0.9462$ and a mean squared error $\Psi = 0.00225$.

Figure 3.9 gives a detailed overview on the calibration results by the squared relative difference $(PS^{\text{G2++}}(m,s)/PS^{\text{market}}(m,s) - 1)^2$.

After having calibrated each of the three interest rate models proposed at the beginning of Sect. 3.5, it is now time for a stock taking. Thus, the reader is shown what has been achieved so far and what still has to be done.

First of all, remember the objective of the current section. Analogously to the share and the property price process as outlined in Sect. 3.3, we are seeking a way to model ZCB prices. Heuristically three different interest rate models were chosen. Firstly, the Cox-Ingersoll-Ross (CIR) model that is a quite tractable construct allowing for positive interest rates only but it does not fit the currently observed term structure of interest rates. Secondly, the Hull-White (HW) model that is a widely

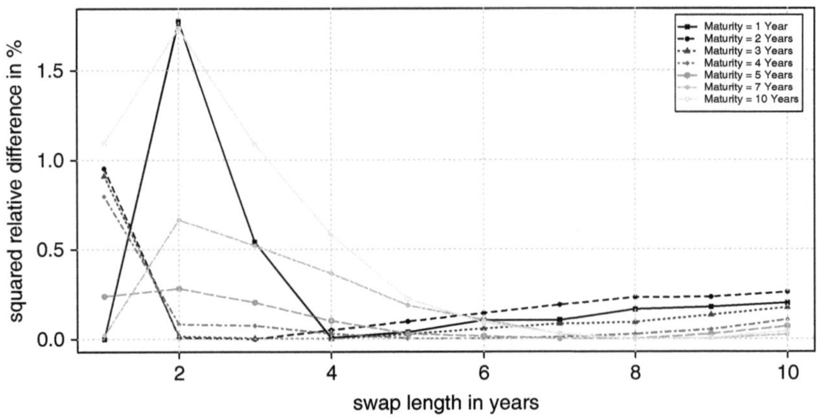

Fig. 3.9 Calibration results for the G2++ model as at the end of 2009 in % given by the squared relative difference $(PS^{\text{G2++}}(m,s)/PS^{\text{market}}(m,s) - 1)^2$

accepted one factor model and thirdly, the two additive factor Gaussian (G2++) model providing the greatest flexibility of all models considered. However, the latter two models are both able to fit the currently observed term structure of interest rates.

Afterwards each of the models was calibrated to observable market instruments, ZCB prices for the CIR model, swaption prices for the HW and the G2++ model. The calibration results are to be discussed in the upcoming section. In the end I will use the calibrated models to generate a set of 5,000 stochastic ZCB prices evolving over 60 years. Below there is an analysis thereof, too. Hereby the design for the proposed modelling of the bond price process is completed.

3.5.4 Comparison of the Interest Rate Model Calibration

This section starts with a summary of the resulting parameters from the calibration procedure presented above. Note that the Ψ value of the CIR model cannot be compared with the respective Ψ value of the other two models, since the instruments underlying the calibration are different. Pricing swaptions with a poor CIR model would not make sense in practice and so the calibration was based on "less sophisticated" instruments like ZCB prices. On the other hand notice that the HW model and the G2++ model perfectly fit the current term structure of interest rates and so these models would perfectly match the set of ZCB prices the CIR model was calibrated to. However keeping that in mind only the Ψ values of the HW model and the G2++ model are comparable. Table 3.7 summarises the calibration results as at the end of 2009 for the CIR, the HW and the G2++ model. Notice that the Ψ value for the CIR model cannot be compare to the Ψ of the HW model or the G2++ model since they were calibrated to different market instruments.

Table 3.7 Summarising the calibration results as at the end of 2009 for the CIR, the HW and the G2++ model

Model	MSE Ψ	Parameters	SDE of the short rate
CIR	0.000189	$k = 0.2150$	$dr(t) = k[\theta - r(t)]dt + \sigma dW(t)$
		$\sigma = 0.1329$	
		$\theta = 0.0411$	
HW	0.00336	$a = 0.0167$	$dr(t) = [\vartheta(t) - ar(t)]dt + \sigma dW(t)$
		$\sigma = 0.0079$	$\vartheta(t) = \frac{\partial f^M(0,t)}{\partial t} + f^M(0,t) + \frac{\sigma^2}{2a}(1 - e^{-2at})$
G2++	0.00225	$a = 2.3660$	$r(t) = x(t) + y(t) + \varphi(t)$
		$b = 0.0327$	$dx(t) = -ax(t)dt + \sigma dW^1(t)$
		$\sigma = 0.0166$	$dy(t) = -by(t)dt + \eta dW^2(t)$
		$\eta = 0.0089$	$dW^1(t)dW^2(t) = \rho dt$
		$\rho = -0.9462$	

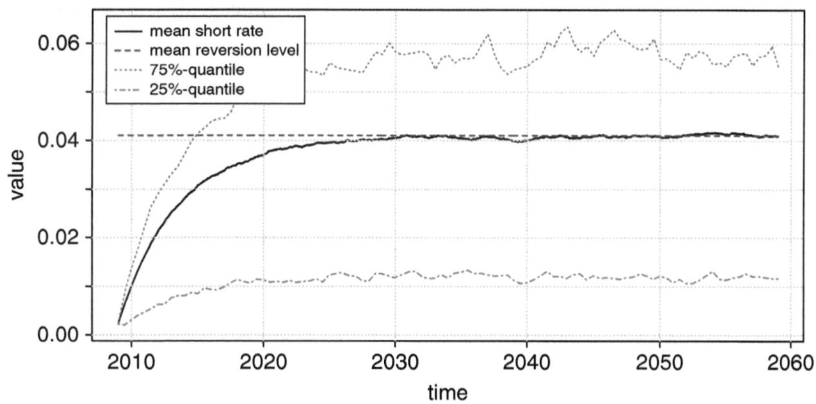

Fig. 3.10 Mean short rate in the CIR model based on 5,000 simulations with parameters $k = 0.2150, \sigma = 0.1329$ and $\theta = 0.0411$

Referring to these parameters as the *best-fit parameters*, only they will be used for the remaining of this section.

Some remarks regarding the parameters are in order. The θ indicates the mean reversion level for the CIR model. This means that the short rate process is pushed towards that level. Figure 3.10 shows the mean short rate in the CIR model based on 5,000 simulations using Formula 3.38 with parameters $k = 0.2150, \sigma = 0.1329$ and $\theta = 0.0411$. The dashed line at 0.411 marks the mean reversion level θ, the upper dotted line and the lower dotted line indicate the 75%-quantile and the 25%-quantile respectively.

The parameter k indicates the speed the current short rate converges to its mean reversion level. We constitute that the distance is halved each $1/k = 4.65$ years.

Considering the two quantiles plotted in the figure we further remark that the short rate is not symmetrically distributed around its mean reversion level. This is caused by the square root term in the SDE of the short rates given by Eq. 3.38. Values above the mean reversion level are pressed down faster than in the opposite case.

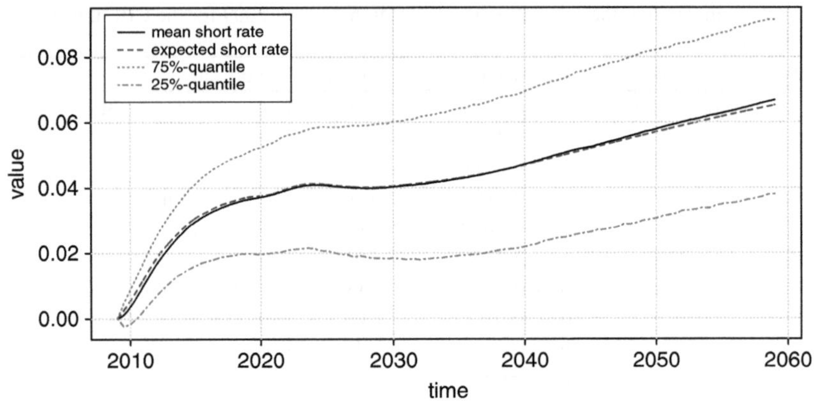

Fig. 3.11 Mean short rate in the HW model based on 5,000 simulations with parameters $a = 0.0167$ and $\sigma = 0.0079$

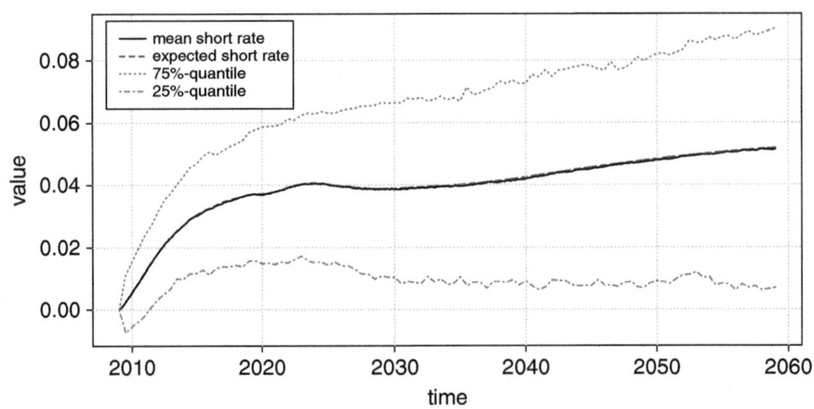

Fig. 3.12 Mean short rate in the G2++ model based on 5,000 simulations with parameters $a = 2.3660, b = 0.0327, \sigma = 0.0166, \eta = 0.0089$ and $\rho = -0.9462$

Above are given the respective Figs. 3.11 and 3.12 for the HW model and the G2++ model. Using Formula 3.45 for the HW model and Formula 3.56 for the G2++ model, again 5,000 simulation runs were generated. Since for these models the short does not converge towards a static mean reversion level, the plots contain the unconditional expected short rate $\mathbb{E}[r(t)] = \mathbb{E}[r(t)|\mathcal{F}_0]$ that can be derived from Formula 3.47 and 3.61 for the HW model and the G2++ model respectively.

For both models the mean short rate fits its unconditional expectation quite well and unlike the CIR model the short rate lies symmetrically around the expected value. Given the best-fit parameters the HW model seems to behave quite similar as the G2++ for the first 20 years of simulation, where $T = 20$ corresponds to

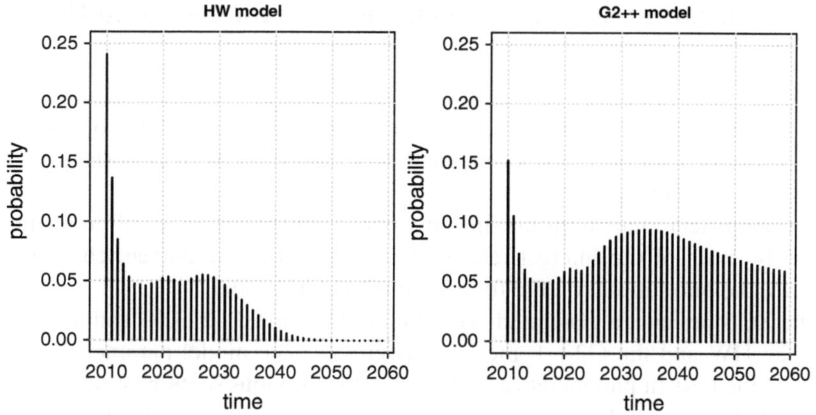

Fig. 3.13 Probability of negative interest rates in the HW model and the G2++ model with parameters $a = 0.0167$ and $\sigma = 0.0079$ (HW) and $a = 2.3660, b = 0.0327, \sigma = 0.0166, \eta = 0.0089$ and $\rho = -0.9462$ (G2++)

the maximum duration of the data basis used for calibration purposes.[15] For $t > T$ the mean short rates start to diverge whereas the HW model generates higher values for the short rate in the mean than the G2++ model. However, in the end, we are interested in ZCB prices and not in the short rate itself. Thus we do not need to further discuss these values but focus on the simulated ZCB prices being presented below.

As discussed above a drawback of the HW model and the G2++ model are the possibility of negative interest rates. As seen by Formula 3.49 and 3.63, the probabilities can explicitly be determined. The reader might consider Fig. 3.13 to get an impression of the respective probabilities given the best-fit parameters.

These plots show that the probabilities of negative interest rates are effectively not marginal at all. Within the first 12–15 years both models produce approximately the same probabilities around 5% before the different trends of the two models appear.

If we allow for negative interest rates this can significantly increase the values of the embedded options like, for example, the interest rate guarantee. But notice that negative interest rates can actually be observed occasionally at markets and so these models are not a priori deemed to be inappropriate. For further information on the history of negative interest rates I refer to the recently published discussion paper Menner and Ilgmann (2011).

In the end it is up to the user whether to accept the probability of negative interest rates or not. However, bearing in mind possible consequences due to negative interest rates, I will nevertheless make use of the HW model and the G2++ model for modelling purposes.

[15]Confer Table A.2 in Appendix A.1.

3.6 Generating Stochastic Scenario Files for the Bond Price Process

Same as for the share and the property price process a set of 5,000 stochastic ZCB prices evolving over 60 years is generated. Here one simulation consists of six ZCB prices of different maturities $T_i \in \{1, 3, 5, 10, 15, 20\}$.

The empirical analysis following in Chap. 5 will investigate different impacts caused by each of the interest rate model presented in the current chapter. Thus there will be one simulation file for each interest rate model.

The ZCB prices are generated according to Formula 3.42, 3.50 and 3.64 for the CIR, the HW and the G2++ model respectively. The model parameters shall be given by the best-fit parameters as outlined in the previous section. Where required, market ZCB prices and market forward interest rates were derived from the currently observed swap rates as at the end of 2009 as described on page 57.

See Table 3.8 for a snapshot of a stochastic scenario file for ZCB prices generated with the CIR model. Naturally the first column contains the observed market data as at the end of 2009 followed by simulated ZCB prices for the subsequent years.

Table 3.8 Snapshot of a stochastic scenario file for ZCB prices

			Year						
			2009	2010	2011	2012	2013	2014	...
Maturities	Simulation 1	1	0.9947	0.9772	0.9471	0.9512	0.9329	0.9777	...
		3	0.9651	0.9251	0.8577	0.8666	0.8268	0.9263	...
		5	0.9181	0.8695	0.7849	0.7959	0.7469	0.8711	...
		10	0.7788	0.7349	0.6452	0.6567	0.6057	0.7366	...
		15	0.6516	0.6172	0.5381	0.5483	0.5035	0.6187	...
		20	0.5523	0.5176	0.4505	0.4591	0.4212	0.5188	...
	Simulation 2	1	0.9947	0.9931	0.9825	0.9642	0.9598	0.9775	...
		3	0.9651	0.9618	0.9372	0.8957	0.8856	0.9258	...
		5	0.9181	0.9165	0.8851	0.8323	0.8197	0.8704	...
		10	0.7788	0.7858	0.7517	0.6952	0.6818	0.7359	...
		15	0.6516	0.6623	0.632	0.5821	0.5703	0.6181	...
		20	0.5523	0.5559	0.5301	0.4878	0.4778	0.5183	...
		...	...	...	...	...	...	...	...

In the following I will present a couple of figures to illustrate the behaviour of the ZCB prices for each stochastic scenario file. Starting with Fig. 3.14, we see the mean prices of a 10 year ZCB over time. For all models the mean price drops down from 0.7788 to a level around 0.7 within the first 10 years. Afterwards the mean prices develop in different directions depending on the underlying model. Considering the evolution of the mean short rate for each model the respective evolution of the mean

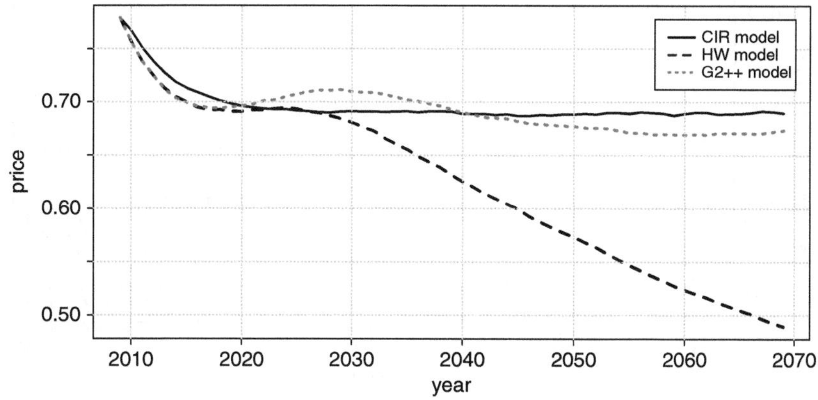

Fig. 3.14 Mean 10 year ZCB price in the CIR, the HW and the G2++ model based on 5,000 simulation runs

10 year ZCB price should not be further surprising.[16] The higher the short rate, the lower the ZCB price and vice versa.

How should one interpret the evolution of the mean price in the HW model? At first sight the continuously decreasing value looks somewhat peculiar. But again this is what could have been expected after having considered the mean short rate in Fig. 3.11. Finally, that evolution is a result caused of the given model parameters and the input data and so it can be seen as a characteristic of the HW model that has to be kept in mind.

Going ahead we see the mean annualised return on the simulated ZCB prices in Figs. 3.15, 3.16 and 3.17. Considering the three plots we observe a steady (CIR) or at least a partial (HW and G2++) inversion of the yield curve. This means that the annualised return on a ZCB with a higher maturity is lower than the annualised return on a ZCB with a lower maturity. Indeed this could have been observed a couple of times in the past but it is still a rare phenomenon. In that sense it is a clear drawback of the CIR model.

The mean annualised returns lead immediately to the future yield curves, as plotted in Figs. 3.18, 3.19 and 3.20. Yield curves with negative slopes are equivalent to the inverted yield curves mentioned before. A constant or flat yield curve means that annualised returns are independent of the bond maturity.

[16]Confer Figs. 3.10, 3.11 and 3.12.

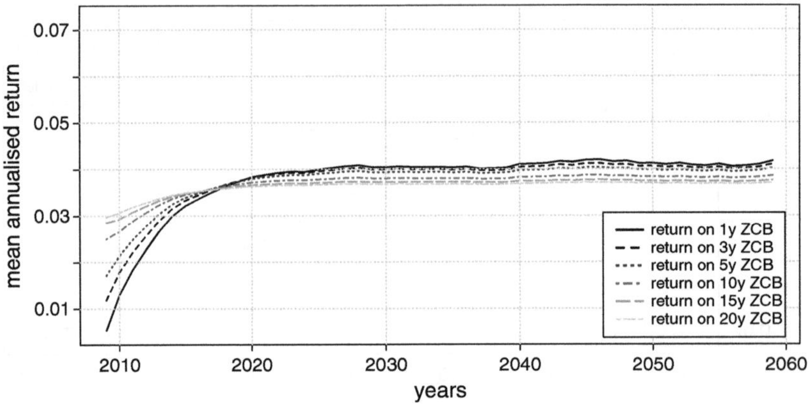

Fig. 3.15 Mean annualised returns on ZCBs in the CIR model based on 5,000 simulation runs

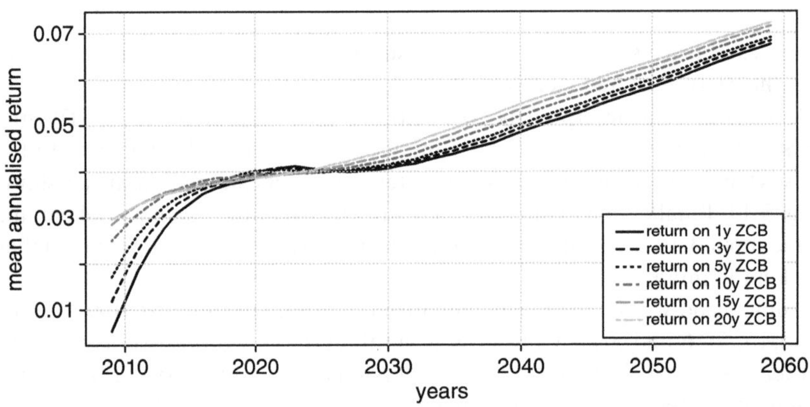

Fig. 3.16 Mean annualised returns on ZCBs in the HW model based on 5,000 simulation runs

After having analysed the different stochastic scenario files a brief conclusion is in order. By introducing the theoretical background of the three interest rate models, its drawbacks like probable negative interest rates or non-fitting of the currently observed term structure of interest rates were mentioned. Thus there was no interest rate model outshining the others, since all had at least one feature that we had to criticise. However, then each of the interest rate models was calibrated to market data and the calibration results were discussed, followed by the generation of the stochastic scenario files for the bond price process.

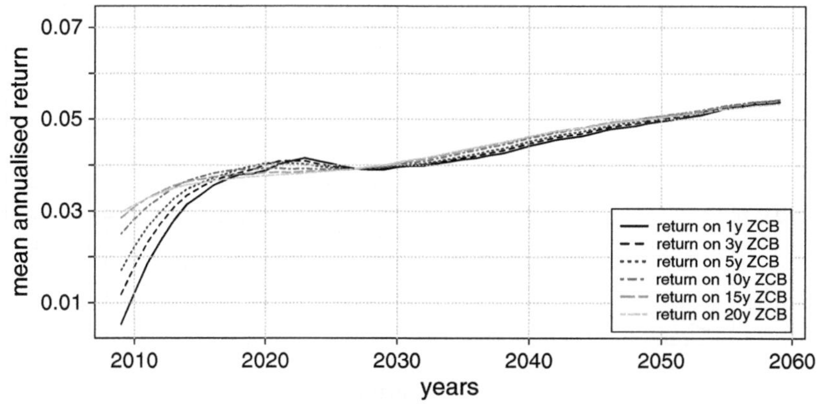

Fig. 3.17 Mean annualised returns on ZCBs in the G2++ model based on 5,000 simulation runs

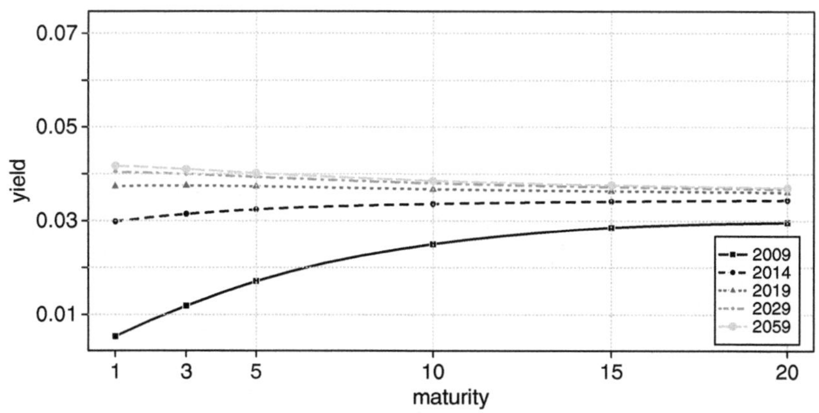

Fig. 3.18 Expected future yield curve in the CIR model based on 5,000 simulation runs

In the end the stochastic ZCB prices are what we have been looking for. Hence its analysis as presented in the current section should turn the balance regarding an appropriate modelling of ZCBs and similar assets. Surely it is debatable whether to accept negative interest rates, inverted yield curves or steady increasing interest rates in the long run or not. But after all, it is still up to the reader in how far he or she trusts a particular interest rate model. So bearing in mind the insights gained on the three interest rate models presented in that chapter, each of them will be applied within the empirical analysis.

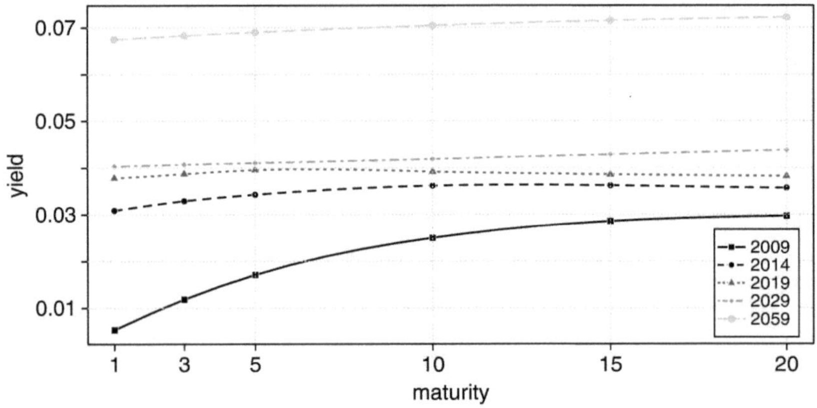

Fig. 3.19 Expected future yield curve in the HW model based on 5,000 simulation runs

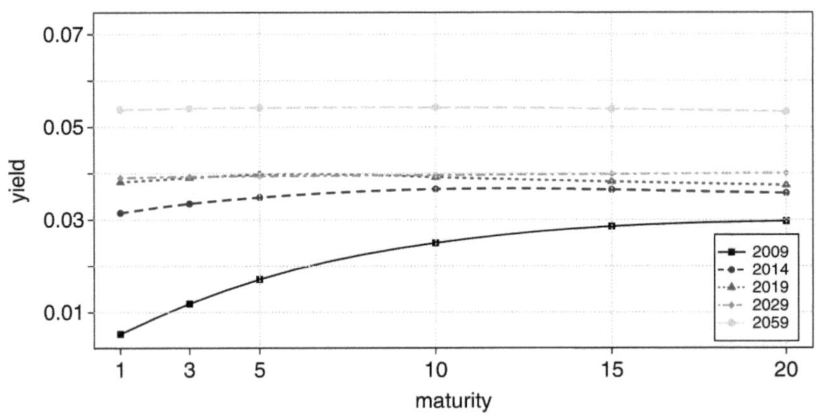

Fig. 3.20 Expected future yield curve in the G2++ model based on 5,000 simulation runs

3.7 Using the Stochastic Scenario Files for Modelling Purposes

In order to complete this chapter there is an example to illustrate the reader the use of the stochastic scenario files for modelling purposes.

Given the model's underlying balance sheet as denoted in Table 3.1 its assets A shall be invested at the financial market. Suppose that market to be perfect where transaction costs, liquidity squeezes, tax effects and other market imperfections may be ignored. The market provides different investment opportunities consisting of an equity fund, a property fund and ZCBs of maturities $T \in \{1, 3, 5, 10, 15, 20\}$ years. Its prices are given by the respective stochastic scenario files as they have been introduced in the current chapter. At this market it shall be possible to buy or sell any share of each asset at any point in time.

Before starting a simulation run I define by $\mathcal{A} \in \{\mathbb{R}^n_+ \mid ||\mathcal{A}||_1 = 1\}$ an asset allocation that specifies the weights of each financial instrument relative to the total market value of the amount invested. It is part of the investment strategy to keep that weights constant over the whole simulation cycle. Hence at the end of each simulation period we are able to buy or sell any unity of each of the instruments at the listed price in order to reallocate the portfolio according to the defined asset allocation.

Now assume an initial amount of money $A_0 = 100$ and an asset allocation $\mathcal{A} = (0.20, 0.45, 0.10, 0.25)$ meaning that we initially buy the following portfolio that is called the *reference portfolio*. The quantities are calculated using the today's market prices as presented below in Table 3.9.

Table 3.9 Sample initial reference portfolio

Asset	Weight	Market value	Quantity
Equity fund	0.20	20	20
Property fund	0.45	45	45
1 year ZCB	0.10	10	10.05
3 year ZCB	0.25	25	25.90
Total	1.00	100	

From the stochastic scenario files we take one simulation path for each of the instruments whereas the time period has a length of 1 year. Therefor see Table 3.10.

Table 3.10 Sample stochastic scenarios. (∗) 2 year ZCB prices are linearly interpolated for this example

Asset	Prices at time					
	$t = 0$	$t = 1$	$t = 2$	$t = 3$	$t = 4$	$t = 5$
Equity fund	1	0.9256	1.0192	0.8206	0.8721	0.5685
Property fund	1	1.1156	1.1582	1.2018	1.3197	1.4338
1 year ZCB	0.9947	0.9868	0.9749	0.9684	0.9660	0.9614
2 year ZCB (∗)	0.9799	0.9656	0.9446	0.9334	0.9300	0.9224
3 year ZCB	0.9651	0.9443	0.9143	0.8983	0.8940	0.8835

After 1 year of simulation the market values of each asset has changed and thus also the market value of the whole portfolio. The current price of the funds can be derived immediately from the stochastic scenario files.

$$\begin{aligned}\text{Value Equity Fund}_{t=1} &= \text{Price Equity Fund}_{t=1} \cdot \text{Quantity Equity Fund}_{t=0}\\ &= 0.9256 \cdot 20 = 18.51\end{aligned}$$

$$\begin{aligned}\text{Value Property Fund}_{t=1} &= \text{Price Property Fund}_{t=1} \cdot \text{Quantity Property Fund}_{t=0}\\ &= 1.1156 \cdot 45 = 50.20\end{aligned}$$

The values of the bonds are more sophisticated to calculate. After the first year the 1 year ZCB has matured and we receive the face value equal to one times the quantity. The 3 year bond becomes a 2 year ZCB and thus we assume the market value to be equal to the price of the 2 year ZCB in $t = 1$.

$$\begin{aligned} \text{Value 3 y ZCB}_{t=1} &= \text{Price 2 y ZCB}_{t=1} \cdot \text{Quantity 3 y ZCB}_{t=0} \\ &= 0.9656 \cdot 25.90 = 25.01 \end{aligned}$$

Hence at the end of the first year before any reallocation the portfolio looks as shown in Table 3.11.

Table 3.11 Sample reference portfolio in $t = 1$ before reallocation

Asset	Weight	Market value	Quantity
Equity fund	0.18	18.51	20
Property fund	0.48	50.20	45
Cash	0.10	10.05	–
1 year ZCB	0.00	0.00	0.00
3 year ZCB (2 years remaining)	0.24	25.01	25.90
Total	1.00	103.77	

We observe that the weights of the assets have changed due to the change in their market values. Thus in order to reallocate the asset allocation we need to buy some equities, 1 year ZCBs and 3 year ZCBs. Notice that to determine the weight of the 3 year ZCB the current total market value of all ZCBs having had an initial maturity of 3 years is divided by the market value of the entire portfolio. For ZCBs of longer maturities the weights are determined analogously. The assets needed to buy are financed by the amount of cash and the part of the property fund we have to sell. After reallocation we have the reference portfolio as shown in Table 3.12.

Table 3.12 Sample reference portfolio in $t = 1$ after reallocation

Asset	Weight	Market value	Quantity
Equity fund	0.20	20.76	22.42
Property fund	0.45	46.70	41.86
Cash	0.00	0.00	–
1 year ZCB	0.10	10.38	10.52
3 year ZCB (2 years remaining)	0.24	25.01	25.90
3 year ZCB (3 years remaining)	0.01	0.93	0.99
Total	1.00	103.77	

We go one step ahead and consider the portfolio at the end of the second year before reallocation. The respective market values shown in Table 3.13 are calculated as for the recent year.

Table 3.13 Sample reference portfolio in $t = 2$ before reallocation

Asset	Weight	Market value	Quantity
Equity fund	0.21	22.85	22.42
Property fund	0.45	48.48	41.86
Cash	0.10	10.52	–
1 year ZCB	0.00	0.00	0.00
3 year ZCB (1 year remaining)	0.23	25.25	25.90
3 year ZCB (2 years remaining)	0.01	0.93	0.99
Total	1.00	108.04	

In Table 3.14 the portfolio is again reallocated.

Table 3.14 Sample reference portfolio in $t = 2$ after reallocation

Asset	Weight	Market value	Quantity
Equity fund	0.20	21.61	21.20
Property fund	0.45	48.62	41.98
Cash	0.00	0.00	–
1 year ZCB	0.10	10.80	11.08
3 year ZCB (1 year remaining)	0.23	25.25	25.90
3 year ZCB (2 years remaining)	0.01	0.93	0.99
3 year ZCB (3 years remaining)	0.01	0.82	0.90
Total	1.00	108.04	

And so the process goes on. I will terminate that example by simulating another year to demonstrate a situation where more than just one bond matures. Therefor look at Table 3.15.

Table 3.15 Sample reference portfolio in $t = 3$ before reallocation

Asset	Weight	Market value	Quantity
Equity fund	0.16	17.40	21.20
Property fund	0.47	50.45	41.98
Cash	0.35	36.99	–
1 year ZCB	0.00	0.00	0.00
3 year ZCB (1 year remaining)	0.01	0.96	0.99
3 year ZCB (2 years remaining)	0.01	0.84	0.90
Total	1.00	106.63	

And finally the reference portfolio at the end of the third year after reallocation looks as shown in Table 3.16.

Table 3.16 Sample reference portfolio in $t = 3$ after reallocation

Asset	Weight	Market value	Quantity
Equity fund	0.20	21.33	25.99
Property fund	0.45	47.98	39.93
Cash	0.00	0.00	–
1 year ZCB	0.10	10.66	11.01
3 year ZCB (1 year remaining)	0.01	0.96	0.99
3 year ZCB (2 years remaining)	0.01	0.84	0.90
3 year ZCB (3 years remaining)	0.23	24.86	27.67
Total	1.00	106.63	

The attentive reader might have observed that we never had to sell any ZCBs. Since there is usually more than one ZCB of the same initial maturity it is not that obvious how to sell them. In such a situation all the ZCBs of a particular (initial) maturity are sold proportionally.

Given any asset allocation one can consider the path of the portfolio's total market value, before or after reallocation, as the price path of one financial entity.

In the next chapter the modelling of the liability side is presented. By doing so the implementation of the asset modelling process into the entire modelling framework can finally be specified.

Chapter 4
Liability Modelling Process

In this chapter I will complete the presentation of the entire modelling approach by introducing its second part, the liability modelling process. As mentioned at the beginning of Chap. 3 insurance specific data is provided by Helvetia. In order to understand the ideas of the liability modelling process certain knowledge of these data is quite helpful. So first of all we shall have a closer look at the available 111,437 policies captured as at the end of 2009. Basically the data can be separated into four categories of different insurance types: Endowment, whole life, term life and pure endowment insurance.

Within each category C we find multiple portfolios of policies $\mathcal{P}(C, \text{TG}_i)$, each based on a particular tariff generation TG_i. A tariff generation characterises the basis for actuarial calculations used to specify the contracts. Typically a tariff generation is denoted by the year it was introduced, so for example TG1980 stands for the tariff generation introduced in 1980, TG2003 for the tariff generation introduced in 2003 and so on. See Fig. 4.1 for an illustrative structuring of the endowment insurance policies.

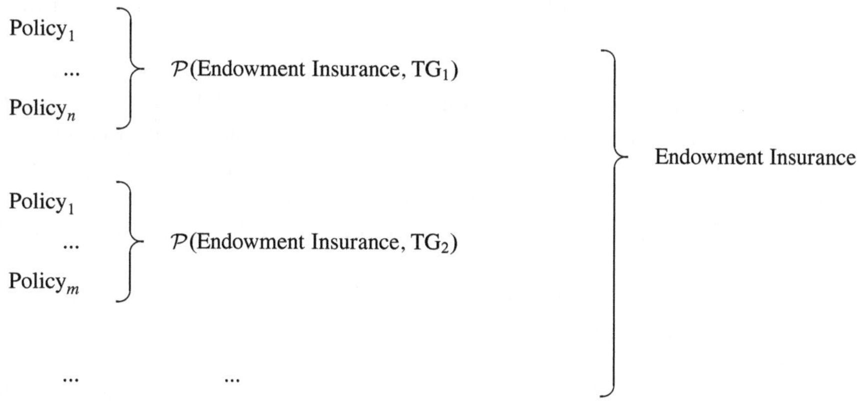

Fig. 4.1 Structuring the endowment insurance policies

N. Rüfenacht, *Implicit Embedded Options in Life Insurance Contracts*,
Contributions to Management Science, DOI 10.1007/978-3-7908-2843-6_4,
© Springer-Verlag Berlin Heidelberg 2012

Pure endowment policies are sparely sold in Switzerland nowadays. Hence it should no longer be surprising that there are only two portfolios of pure endowment policies for TG1980 and TG1994 within the data.

In contrast to this we have endowment policies for TG1970, TG1980, TG1994, TG1996, TG1999 and TG2003 providing approximately 86.5% of all data available. Finally we will not further discuss the specifications of different tariff generations itself but pick that issue up again within the empirical analysis following in Chap. 5.

Below Table 4.1 provides an overview on the key quantities for the whole data basis, prepared for each of the four categories. Quantities of time are denoted in years, amounts of money are denoted in Swiss Francs. The value in brackets stand for the mean values per policy.

4.1 The Initial Values of the Balance Sheet Entries

Remember the model's underlying balance sheet pictured in Table 3.1. We have seen how to model the assets A over time when defining the asset modelling process. But I have neither defined its initial value $A(0) = A_0$ nor put its evolution into the context of the entire modelling approach. This shall be handled in the subsequent.

To define a reasonable starting value for these assets I simply ask for *the amount we effectively have*. Supposing that Helvetia is a well financed company we assume that all policy reserves as well as all surplus reserves disclosed in the data are actually covered by some real existing assets. Thus for any particular portfolio $\mathcal{P}$ that amount is assumed to be existing and I hence define

$$A_0 = \text{POL\_RES}(\mathcal{P}) + \text{SURP\_RES}(\mathcal{P}), \tag{4.1}$$

with $\text{POL\_RES}(\mathcal{P})$ and $\text{SURP\_RES}(\mathcal{P})$ the total amount of policy reserves of the portfolio $\mathcal{P}$ and the total amount of surplus reserves of the portfolio $\mathcal{P}$ respectively.

Note that by making this assumption we transform a statutory quantity into a market consistent quantity. Even though POL_RES and SURP_RES are statutory quantities there must be some tangible assets of the same amount to cover them. And it will be exactly this amount A_0 to be invested in the reference portfolio as explained in Sect. 3.7 and thus it becomes a market consistent value. However, the empirical analysis will show that there are also difficulties arising from this transformation.

On the other side of the balance sheet there is, amongst others, the policy account P. This position represents *the amount we effectively need*. Again we are seeking a plausible and market consistent value for this.

Given any portfolio $\mathcal{P}$, no matter what kind of insurance type, one can project its future cash flows using best estimate assumptions. Discounting the future cash outflows less the future cash inflows with the current market interest rate curve

Table 4.1 Overview on the available 2009 policy data. Quantities of time are denoted in years, amounts of money are denoted in Swiss Francs. SP/RP stands for insurance paid against single premium/regular premium. The values in brackets are the mean values per policy

	Endowment		Whole life		Term life	Pure endowment		Total
	SP	RP	SP	RP	RP	SP	RP	
Number of policies	24,112	72,246	1,045	930	11,181	550	1,373	111,437
Mean policy term	21.04	25.25	(**)	(**)	23.30	23.78	21.26	24.06
Mean age at entry	42.49	37.28	61.45	53.68	37.20	36.15	41.48	38.81
Mean duration in force	11.88	9.78	16.01	17.74	10.65	15.70	10.43	10.48
Tariff premium (SP)	838,163,128 (34,761)		31,559,984 (30,201)			11,796,203 (21,448)		881,519,315 (34,291)
Tariff premium (RP) (*)		4,078,744,854 (2,286)		(***)	229,464,917 (980)		56,887,168 (1,745)	4,365,096,940 (2,100)
Sum assured	1,134,699,062 (47,059)	4,499,945,012 (62,286)	48,574,321 (46,482)	19,266,095 (20,716)	1,344,726,761 (120,269)	22,191,358 (40,348)	81,062,865 (59,041)	7,150,465,474 (64,166)
Surplus reserve	125,782,928 (5,216)	141,793,552 (1,963)	15,950,028 (15,263)	1,206,337 (1,297)	9,588,529 (858)	3,406,066 (6,193)	1,196,874 (871)	298,924,314 (2,682)
Policy reserve (SP)	956,711,548 (39,678)		37,691,327 (36,068)			18,122,367 (32,950)		1,012,525,242 (39,387)
Policy reserve (RP)		1,417,156,537 (19,616)		10,574,397 (11,370)	24,854,309 (2,223)		36,509,154 (26,591)	1,489,094,397 (17,370)

(*) Yearly tariff premium multiplied by the premium payment duration
(**) Whole life insurance do not have a fixed policy term
(***) Since whole life insurance do not have a fixed policy term the sum of premiums cannot be determined

yields the requested amount we effectively need for that portfolio $\mathcal{P}$ at start, P_0. This sounds quite obvious and indeed it is, at least in theory. For practical reasons some further explanations are in order.

A core element of this liability modelling process is a best estimate cash flow scheme representing an ordinary runoff of any insurance portfolio. Best estimate assumptions presuppose mortality rates and cost parameters to be calculated without any margin and to remain unchanged whilst runoff. Cash flows are determined on a policy level and aggregated to a portfolio level by summing up policy values for every period of time.

Before specifying the particular cash flow generating quantities some auxiliary quantity has to be defined: The *in-force probability* (ifp). Basically the ifp is the today's survival probability of the policyholder. Once the policy has matured the ifp is zero. Matured policies produce no more cash flows and are no longer of interest. So to come up to the best estimate assumption all cash flows have to be weighted with that probability.

Given an x year old policyholder in $t = 0$. Its policy was issued some time ago and will mature in $t = T$. At the beginning of any simulation ($t = 0$) the value of the $\text{ifp}_0 = 1$. One period later

$$\text{ifp}_1 = \text{ifp}_0 \cdot (1 - {}_{\frac{1}{N}}q_{\lfloor x \rfloor}) \tag{4.2}$$

where $(1 - {}_{\frac{1}{N}}q_x)$ is the $\frac{1}{N}$ year survival probability of an x year old person. Note that we have implicitly assumed that the period has a length of $\frac{1}{N}$ years where e.g., $N = 12$.

The mortality rates q_x being applied for the purpose of this thesis will be taken from the mortality table *EKM/F_01_05_II* that is also used by Helvetia. EKM/F_01_05_II denotes a second order mortality table containing the 1 year mortality rates for men and women of ages 0–116 (men) and 118 (women) developed by the *Subkommission "Statistik"*, a working group of the *Swiss Insurance Association* (SIA). A *second order* mortality table assumes expected mortality rates without any margins. The table's underlying data basis was collected from capital forming insurance within an observation period from 2001 to 2005. Since all of the data I am working with belongs to the class of capital forming insurance this mortality table may be considered as an appropriate one.

As in the case of the EKM/F_01_05_II mortality rates q_x are usually supplied for integer ages only and so the floor of the current age $\lfloor x \rfloor$ is taken. But how can mortality rates be derived for periods of less than a year? Obviously we need ${}_{\frac{1}{N}}q_x$ such that

$$q_x = \sum_{n=1}^{N} (1 - {}_{\frac{1}{N}}q_x)^{n-1} \, {}_{\frac{1}{N}}q_x \qquad \forall\, x \geq 0,\, N > 0. \tag{4.3}$$

Theorem 1. *For any* $x \geq 0,\ N > 0,$

$$_{\frac{1}{N}}q_x := 1 - (1 - q_x)^{1/N} \tag{4.4}$$

fulfils Eq. 4.3.

Proof.

$$\begin{aligned}
q_x &= \sum_{n=1}^{N} (1 - {_{\frac{1}{N}}q_x})^{n-1}\ {_{\frac{1}{N}}q_x} \\
&= \sum_{n=1}^{N} \left(1 - (1 - (1 - q_x)^{1/N})\right)^{n-1} (1 - (1 - q_x)^{1/N}) \\
&= \sum_{n=1}^{N} (1 - q_x)^{(n-1)/N} (1 - (1 - q_x)^{1/N}) \\
&= \sum_{n=1}^{N} (1 - q_x)^{(n-1)/N} - (1 - q_x)^{n/N} \\
&= (1 - q_x)^{0/N} - (1 - q_x)^{N/N} \\
&= 1 - (1 - q_x) \\
&= q_x
\end{aligned}$$

□

Finally the ifp for any t^*, $0 < t^* \leq T$, with T indicating the time of maturity, is as follows

$$\text{ifp}_{t^*} = \text{ifp}_0 \cdot \prod_{t=1}^{t^*} (1 - {_{\frac{1}{N}}q_{\lfloor x + \frac{t-1}{N} \rfloor}}) \qquad \forall\, 0 < t^* \leq T \tag{4.5}$$

where ${_{\frac{1}{N}}q_{\lfloor x + \frac{t-1}{N} \rfloor}}$ is defined as in Formula 4.4.

One period after having matured the policy doesn't exist any longer and thus its ifp turns to zero.

$$\text{ifp}_{T + \frac{t}{N}} = 0 \qquad \forall t > 0$$

Example

Given a 57 years and 9 months old man holding any kind of insurance policy. We want to calculate the ifp for the upcoming 6 months when the policy matures at the end. Let $q_{57} = 0.004644$ and $q_{58} = 0.005092$. So according to Formula 4.4 we know that

$$_{\frac{1}{12}}q_{57} = 1-(1-q_{57})^{1/12} = 0.000388$$

and

$$_{\frac{1}{12}}q_{58} = 1-(1-q_{58})^{1/12} = 0.000425$$

Taking Formula 4.5 the ifp can easily be calculated then. Find the calculations in Table 4.2.

Table 4.2 Illustrative example for the ifp

x	57+9/12	57+10/12	57+11/12	58	58+1/12	58+2/12	58+3/12	58 + 4/12
$\lfloor x \rfloor$	57	57	57	58	58	58	58	
$q_{\lfloor x \rfloor}$	0.004644	0.004644	0.004644	0.005092	0.005092	0.005092	0.005092	
$_{\frac{1}{12}}q_{\lfloor x \rfloor}$	0.000388	0.000388	0.000388	0.000425	0.000425	0.000425	0.000425	
ifp	1	0.999612	0.999225	0.998837	0.998412	0.997987	0.997563	0

After 6 months the policyholder's policy is in-force with probability 0.997563 at maturity. This is equivalent to the probability of the policyholder to be alive after 6 months. After having paid possible maturity benefits the policy has matured and the ifp turns to 0 for the following period.

Recall the objective of the current section. We are still on the way to define a reasonable starting value for the policy account $P_0^{\mathcal{P}}$. Having said that this value shall be quantified by the discounted future cash outflows less the discounted future cash inflows, I will now proceed by specifying what outflows and inflows I am talking about. Again I emphasise that cash flow calculations are made on best estimate assumptions.

Premium Income (PREM_INC)

Premiums are to be paid in advance and consist of tariff premium net of any deductions. It is the net amount the policyholder has to pay, once in case of a single premium policy or multiple times in case of a regular premium policy.

Regular premiums can become due monthly, quarterly, half-yearly or yearly depending on the policyholder's taste. The premium payment duration is less or equal to the policy term. Finally the premium income of a period is the premium to be paid weighted with the ifp.

Expenses (EXPENSES)

Expenses are an aggregation of variable and fixed expenses. Variable expenses are calculated subject to a given quantity, usually the sum assured or the sum of premiums. Therefore variable expenses subject to premiums are due at the same time as premiums, i.e., in advance and only if premiums are due too, whereas expenses that are subject to the sum assured will be due in advance at the beginning of every period.

Fixed expenses shall also be due in advance. They are defined at a higher level and thus independent from individual policy parameters. There is a yearly amount allocated on a monthly basis to every policy. The total period expenses is the sum of variable and fixed expenses weighted with the ifp.

Death Outgo (DTH_OUT)

Since we know the ifp we can easily calculate the mortality d_t within one period as

$$d_t = \mathrm{ifp}_{t-1} - \mathrm{ifp}_t, \qquad 0 < t \leq T.$$

If the insurance type allows for a death benefit then the death outgo of a period is the sum assured plus a possible surplus weighted with the respective mortality paid in arrear. Otherwise it is zero.

Note that *a possible surplus* depends on the currently disclosed surplus reserve only. That reserve contains earlier allocated surpluses and belongs to the policyholder. Hence it needs to be treated like a guaranteed benefit. Future surpluses are going to be modelled separately so they do not affect the current cash flows.

Maturity Outgo (MAT_OUT)

If the insurance type allows for a maturity benefit the maturity outgo is the sum assured plus a possible surplus weighted with the ifp at maturity T, ifp_T. Else the maturity outgo is zero. Again this outgo is paid in arrear.

Besides, possible surpluses are treated the same as for the death outgo.

Now the quantities producing cash inflows and cash outflows are defined. Let us quickly summarise the suggestions.

Premiums are inflows paid in advance, expenses are outflows falling due in advance whereas maturity outgo and death outgo are outflows falling due in arrear. To get best estimate cash flows we further have to weight them with the ifp that is based on the mortality table EKM/F_01_05_II. For our purposes we are mainly interested in the net cash flow NET_CF_t that is defined as follows:

$$\begin{aligned} \text{NET\_CF}_t &= \text{DTH\_OUT}_t \\ &\quad +\text{MAT\_OUT}_t \\ &\quad +\text{EXPENSES}_t \\ &\quad -\text{PREM\_INC}_t \end{aligned} \tag{4.6}$$

(4.7)

Given the market interest rate curve s_t we can finally determine the initial value of the policy account P_0 for a portfolio $\mathcal{P}$ by

$$P_0 := \text{PV}^{\text{mc}}(\text{NET\_CF}) = \sum_{t=0}^{T} \text{NET\_CF}_t \cdot (1 + s_t)^{-t}, \tag{4.8}$$

where the suffix *mc* indicates market consistency due to the market interest rates used for discounting. Below there is an example to illustrate how to determine the $\text{PV}^{\text{mc}}(\text{NET\_CF})$ for a sample endowment policy.

Example

Again we have a 57 years and 9 months old man holding an endowment insurance policy. See the policy specifications in Table 4.3 below.

Table 4.3 Sample endowment policy specifications

Remaining term	6 months
Sum assured	24,000
Yearly premium	1,200
Premium payment	Monthly
Variable expenses	0.1% of sum assured
Fixed expenses	36

Here we are considering a single policy. Repeating the algorithm below for all policies within a portfolio and summing up over t finally yields the portfolio value. According to the explanations above, the NET_CF can be calculated. See an illustration of that example in Table 4.4.

Table 4.4 Illustration of PV^{mc}(NET_CF) calculation

	$t=0$	$t=1$	$t=2$	$t=3$	$t=4$	$t=5$	$t=6$
ifp	1	0.9996	0.9992	0.9988	0.9984	0.9980	0.9976
PREM_INC	100	99.96	99.92	99.88	99.84	99.80	
Fixed EXPENSES	3	2.999	2.998	2.997	2.995	2.994	
Variable EXPENSES	2	1.999	1.998	1.998	1.997	1.996	
DTH_OUT		9.6	9.6	9.6	9.6	9.6	
MAT_OUT							23,942
NET_CF	−95	−85.36	−85.32	−85.29	−85.25	−85.21	23942.4
s_t		0.0052	0.0055	0.0058	0.0061	0.0064	0.007
NET_CF$_t \cdot (1+s_t)^{-t}$	−95	−85.33	−85.25	−85.24	−85.16	−85.16	23914.58
PV^{mc}(NET_CF)	23393.44						

A remark regarding the time index t. In the above example the meaning of t is twofold. $t = i$ is the end of the ith year but on the same time it is the beginning of the year $i + 1$. By this, payments in arrear for year i occur at the same time as payments in advance for year $i + 1$. This assumption just simplifies the illustration of the example.

To discount the future net cash flow the market interest rate is requested. This enables its present value to become market consistent. For the remainder of this thesis I will exclusively refer to the current Swiss swap rate curve as the market interest rates, where *current* is meant to be as at the end of 2009.[1]

We have two more positions on the liability side of which the initial values need to be defined. There is the undistributed surplus U to accumulate allocated but not yet distributed surpluses. Since earlier allocated surpluses are incorporated in the cash flow projection already and possible future surpluses are determined as time goes by, U is initially zero.

$$U_0 = 0 \tag{4.9}$$

Lastly the buffer B is the residual quantity that shall be defined as follows.

$$B_0 = A_0 - P_0 - U_0 = A_0 - P_0 \tag{4.10}$$

[1] See Fig. A.1 in Appendix A.

Now we have defined the initial values for each of the balance sheet entries. Before proceeding with the outline of the modelling algorithm the reader finds a summary of the respective definitions presented in the current section in Table 4.5.

Table 4.5 Summarising the initial values of the balance sheet entries

Quantity	Initial value	Description
Asset portfolio A_0	$\text{POL\_RES}(\mathcal{P}) + \text{SURP\_RES}(\mathcal{P})$	What we effectively have
Policy account P_0	$\text{PV}^{\text{mc}}(\text{NET\_CF})$	Market consistent present value of future net cash flow. What we effectively need
Undistr. surplus U_0	0	Account for accumulating future surpluses. Initially that account is zero
Buffer B_0	$A_0 - P_0$	Residual quantity

4.2 The Entire Modelling Algorithm

After having defined the asset modelling process in Chap. 3 and the initial values for each entry of the model's underlying balance sheet in the previous section it is time to present the modelling algorithm that finally explains the entire model.

Throughout this section we assume consideration of a portfolio $\mathcal{P}$ of endowment policies whereas its tariff generation is nonessential. Generalisations regarding other insurance types are straight forward.

Now suppose some not further specified asset allocation enabling the generation of a reference portfolio according to the considerations of Sect. 3.7. The time t value of that reference portfolio is denoted by $\text{INDEX}(t)$. Moreover I will take the best estimate cash flow scheme for the portfolio $\mathcal{P}$ as known and use the same terminology as introduced in the previous section.

The entire modelling algorithm consists of the evolution of the following quantities.

- $A(t)$ Time t value of the asset portfolio.
- $P(t)$ Time t value of the policy account.
- $U(t)$ Time t value of the allocated but undistributed surplus.
- $B(t)$ Time t value of the buffer.

Let $[t, t+1)$, $t = 0$ be the first time period to be modelled. The time horizon of the modelling algorithm shall be given by the interval $[0, T+1) = [0, T^+]$, where T^+ denotes the end of the year in which the last policy in the portfolio $\mathcal{P}$ matures. The initial values are set as summarised in Table 4.5, all available money (A_0) is invested in the reference portfolio. At the beginning of the period premiums come in and expenses go out. If the portfolio allows for premium reductions the respective surplus is allocated to the premiums of that period on the asset side and subtracted from the undistributed surplus U on the liability side. Hence we have:

Calculations at the beginning of a period (bp)

$$
\begin{aligned}
A^{bp}(t) &= A^{ab}(t^+ - 1)\\
&\quad +\text{PREM\_INC}(t)\\
&\quad -\text{PREM\_RED}(t)\\
&\quad -\text{EXPENSES}(t)\\
P^{bp}(t) &= P^{ab}(t^+ - 1)\\
&\quad +\text{PREM\_INC}(t)\\
&\quad -\text{EXPENSES}(t)\\
U^{bp}(t) &= U^{ab}(t^+ - 1)\\
&\quad -\text{PREM\_RED}(t)\\
B^{bp}(t) &= A^{bp}(t)\\
&\quad -P^{bp}(t)\\
&\quad -U^{bp}(t)
\end{aligned}
$$

PREM_RED(t) is the amount credited to the policyholder to lower its premium for period $[t, t+1)$, if the portfolio allows for premium reductions. As we will see, that kind of bonus is determined at the end of a period and hence it is always zero when starting a simulation.

The end of a period $[t, t+1)$ is denoted by t^+ and therefore $t^+ - 1$ refers to the end of the foregoing period. In general any kind of bonus is calculated at the end of a period and because of this it is further necessary to declare whether period end calculations are made *before bonus calculations* (bb) or *after bonus calculations* (ab). Above one observes, for example, $P^{ab}(t^+ - 1)$ that denotes the value of the policy reserve P at the end of period $[t-1, t)$ after having calculated a possible bonus.

We go one step forward to the end of period $[t, t+1)$, i.e., to t^+. Earlier allocated bonuses are distributed to the policyholders. At this point in time no new bonuses are calculated yet for the current period and so all the calculations are deemed to be *before bonus* (bb).

The policyholder has the opportunity to chose between two different bonus mechanisms. The first possibility is the reduction of its premiums as mentioned already. The second possibility is an interest-bearing accumulation of all the allocated bonuses. In this case all allocated bonuses are accumulated at a fixed rate i_u and distributed to the policyholder at maturity or death. Whether premium reduction or interest-bearing accumulation will be applied depends on the contractual specification of each single policy and is thus a given setting.

Considering the available data one realises that the vast majority has chosen the interest-bearing accumulation across all types of insurance. To simplify matters I will assume the latter one as the default case for all policies and consider premium reduction in line with a "what-if analysis" in Chap. 5. However to make allowance for that distinction the indicator $\mathbb{I}_{\text{AccB}}$ is 1 if interest-bearing accumulation is designated. Otherwise it is zero.

Another bonus that is independent of the other two is the terminal bonus TERM_BONUS that is paid out at maturity only. Whether a terminal bonus is credited or not is contractually specified over all policies of a particular portfolio. Again there is some indicator $\mathbb{I}_{\text{TB}}$ being equal to 1 if a terminal bonus is designated and zero otherwise.

The difference between an interest-bearing accumulation and a terminal bonus is the different calculation basis. As we see the interest-bearing accumulation depends on the undistributed surplus U. Once any bonus has been allocated to that quantity it is a liability from the insurer's perspective towards its policyholders and it must be distributed at the latest at maturity. On the other hand the amount of the terminal bonus primarily depends on the actual value of the buffer B at maturity. Given that the policy allows for a terminal bonus the policyholder will only receive a bonus if the buffer B is greater than zero at his policy's maturity.[2]

Below there is the itemisation of the calculations at the end of a period, before bonus calculations.

Calculations at the end of a period, before bonus calculations (bb)

$$\begin{aligned}
\text{ACC_BONUS}(t^+) &= \mathbb{I}_{\text{AccB}} \cdot U^{bp}(t)(1+i_u) \\
&\quad \cdot \left(\text{MAT_OUT}(t^+) + \text{DTH_OUT}(t^+)\right) / \\
&\quad \sum \left(\text{MAT_OUT}(s \geq t^+) + \text{DTH_OUT}(s \geq t^+)\right) P(t^+, s) \\
\text{TERM_BONUS}(t^+) &= \mathbb{I}_{\text{TB}} \cdot \min\{\text{max_TB_rate} \cdot \text{MAT_OUT}(t^+), \max\{0, B^{bp}(t)\} \\
&\quad \cdot \text{MAT_OUT}(t^+) / \sum \left(\text{MAT_OUT}(s \geq t^+) P(t^+, s)\right)\} \\
A^{bb}(t^+) &= A^{bp}(t) \cdot \text{INDEX}(t^+)/\text{INDEX}(t) \\
&\quad - \text{MAT_OUT}(t^+) \\
&\quad - \text{DTH_OUT}(t^+)
\end{aligned}$$

[2]The suggested definition here is different from the original one where the amount of the terminal bonus depends on the policy duration. Since the entire model is based on a portfolio cash flow consideration it is no longer possible to identify the individual term of each policy.

$$
\begin{aligned}
& -\text{TERM\_BONUS}(t^+) \\
& -\text{ACC\_BONUS}(t^+) \\
P^{bb}(t^+) &= P^{bp}(t) \cdot (1/P(t,t^+)) \\
& -\text{MAT\_OUT}(t^+) \\
& -\text{DTH\_OUT}(t^+) \\
U^{bb}(t^+) &= U^{bp}(t) \cdot (1 + i_u) \\
& -\text{ACC\_BONUS}(t^+) \\
B^{bb}(t^+) &= A^{bb}(t^+) \\
& -P^{bb}(t^+) \\
& -U^{bb}(t^+)
\end{aligned}
$$

The factors $P(t_1, t_2) = \exp(-(t_2 - t_1) \cdot s(t_1, t_2))$ used to discount are derived from the current swap rate curve, where $s(t_1, t_2)$ denotes the $(t_2 - t_1)$ year swap rate at time t_1.

Let us go ahead and consider the third and last step of the modelling algorithm. The bonus calculations start with the determination of the shareholder's profit SH_PROFIT. The idea behind this is to give the shareholder a fixed percentage of the current end of year policy account, sh_margin $\cdot\, P^{bb}(t^+)$. The profit is financed by the buffer B that actually gives its upper boundary and is paid out immediately.

Consequentially the shareholder's interests are served before the policyholder's interests. However the policyholder's profit, i.e., its bonus, is a variation of the bonus distribution scheme as suggested in Grosen and Jørgensen (2000) with fixed parameters α (distribution ratio) and β (target buffer ratio). Once PH_PROFIT has been determined it will be allocated to the undistributed surplus U.

Lastly, PREM_RED$(t + 1)$, a possible reduction of the premiums for the upcoming period is determined. Here the indicator function can easily be defined as $\mathbb{I}_{\text{PR}} = (1 - \mathbb{I}_{\text{AccB}})$, meaning that a premium reduction is designated if no interest-bearing accumulation is designated and vice versa.

This leads to the following calculations at the end of a period, after bonus calculations.

Calculations at the end of a period, after bonus calculations (ab)

$$
\begin{aligned}
\text{SH\_PROFIT}(t^+) &= \max\{0, \min\{B^{bb}(t^+), \text{sh\_margin} \cdot P^{bb}(t^+)\}\} \\
\text{PH\_PROFIT}(t^+) &= \max\{0, \alpha \cdot (B^{bb}(t^+) - \text{SH\_PROFIT}(t^+) - \beta P^{bb}(t^+))\}
\end{aligned}
$$

$$
\begin{aligned}
U^{ab}(t^+) &= U^{bb}(t^+) \\
&\quad +\text{PH\_PROFIT}(t^+) \\
\text{PREM\_RED}(t+1) &= \mathbb{I}_{\text{PR}} \cdot U^{ab}(t^+) \cdot \text{PREM\_INC}(t+1) \\
&\quad / \sum (\text{PREM\_INC}(s \geq t+1) P(t+1, s)) \\
A^{ab}(t^+) &= A^{bb}(t^+) \\
&\quad -\text{SH\_PROFIT}(t^+) \\
P^{ab}(t^+) &= P^{bb}(t^+) \\
B^{ab}(t^+) &= B^{bb}(t^+) \\
&\quad -\text{SH\_PROFIT}(t^+) \\
&\quad -\text{PH\_PROFIT}(t^+)
\end{aligned}
$$

Herby the modelling algorithm has come full circle and restarts with the calculations at the beginning of the following period $[t+1, t+2)$. The algorithm continues until the last policy in the portfolio has matured in $[T, T+1) = [T, T^+]$. Table 4.6 summarises the key quantities of the modelling algorithm.

Table 4.6 Summarising the key quantities of the entire modelling algorithm

Quantity	Description
$A(t)$	Time t value of the asset portfolio
$P(t)$	Time t value of the policy account
$U(t)$	Time t value of the allocated but undistributed surplus
$B(t)$	Time t value of the buffer
sh_margin	Shareholder margin
α	Distribution ratio
β	Target buffer ratio
i_u	Bonus interest rate
max_TB_rate	Maximum terminal bonus relative to the maturity outgo

This completes the description of the entire modelling algorithm and so we have almost all of the necessary background knowledge to start the empirical analysis. Before doing so I will outline what the relevant outputs of the model will be and refer to the quantitative objectives formulated in Sect. 1.3.

4.3 Using the Model's Output to Determine an Option's Value

Recall the central question I aimed for answering with that model as formulated in Sect. 1.3:

> What value do the considered embedded options and guarantees have, absolutely and relative to the current policy reserves, given a perspective taken from the insurer, the policyholder and the shareholder respectively?

Given that model, I define the today's (absolute) value $V(\cdot)$ of the interest rate guarantee and the dividend option of any considered portfolio as follows.

$$V(\text{interest rate guarantee}) := -\mathbb{E}[B^{ab}(T^+) \cdot (1 + s_{T+1})^{-(T+1)} \mid B(T^+) < 0] \tag{4.11}$$

and

$$\begin{aligned} V(\text{dividend option}) := \mathbb{E}\Big[\sum_{t=0}^{T}\Big(&\text{TERM\_BONUS}(t^+) \\ &+\text{ACC\_BONUS}(t^+) \\ &+\text{PREM\_RED}(t+1)\Big) \cdot (1 + s_{t+1})^{-(t+1)}\Big], \end{aligned} \tag{4.12}$$

with s_t the current $t-$year swap rate.

The value of the interest rate guarantee quantifies the guarantee an insurer promises by accumulating savings premiums with the technical interest rate. With the value of the dividend option we have the present value of the expected future non-guaranteed benefits for the policyholders.

A further quantity that has not been explicitly identified as an embedded option but anyhow exhibits the characteristics thereof from the insurer's point of view is the present value of future shareholder profits, quasi the *shareholder option*.

$$V(\text{shareholder option}) := \mathbb{E}\Big[\sum_{t=0}^{T}(\text{SH\_PROFIT}(t^+)) \cdot (1 + s_{t+1})^{-(t+1)}\Big] \tag{4.13}$$

Similar to the dividend option, the value of shareholder option is the present value of the expected future non-guaranteed benefits for the shareholders.

For the interpretation of the empirical analysis it is essential to understand the exact meaning of the values of these options. Since the insurer holds a short position in these options, he has an obligation towards its policyholders and its shareholders. However the dividend and the shareholder option are in the money if the insurer performs well and realises a profit. Consequently the insurer is only confronted to meet this particular obligation in states where he has already gained that money at some earlier point in time. So in some sense there are, and this is the crucial point, only opportunity costs that occur as the insurer is implicitly hedged against these options. Hence there is no need to set up any reserves for the dividend and the shareholder option today.

Contrariwise the interest rate guarantee is rather a classic option. There is no implicit hedge against this option and so the insurer definitely should set up a reserve today in order to meet future obligations.

As mentioned in the first chapter already, taking someone's perspective with respect to the modelling is equivalent to optimising certain variables under given constraints. On the one hand, the insurer intends to minimise losses in due consideration of competitive dividend and surplus payments. Hence we suggest the insurer minimises the value of the interest rate guarantee, given a targeted value G for the values of the dividend and the shareholder option. One might argue that the insurer could also aim for maximising the buffer B since the model considers its terminal value as some kind of free surplus for the benefit of the company. Indeed he could but when doing so the insurer obviously neglects the interests of its policyholders as well as the interests of its shareholders that would be an unrealistic scenario in practice.

On the other hand the policyholder is only interested in its guaranteed benefits and maximised surpluses. This means that he wants to have a maximised value of the dividend option besides its guaranteed benefits. Lastly the shareholder of course wants to have maximised and stable dividend payments or, in terms of the model, a maximised value of the shareholder option.

Below Table 4.7 summarises the model's output variables to be optimised from each of the different perspectives we shall take.

Table 4.7 Summarising the model's output variables to be optimised

Perspective	Target	Constraint
Insurer	min{interest rate guarantee}	{value of dividend option + value of shareholder option} $= G$
Policyholder	max{value of dividend option}	
Shareholder	max{value of shareholder option}	

4.4 A Sample Simulation

This chapter ends up with an illustrative example of the entire modelling approach. Therefore suppose a sample portfolio $\mathcal{P}^s$ consisting of 50 pure endowment policies maturing within the next 6 years. The bonus scheme being applied is an interest-bearing accumulation and a terminal bonus is provided, too. From the policy data the following cash flow scheme can be derived. The present value of the net cash flow has been calculated using the current swap rate curve assuming a discrete discounting. See the cash flow scheme for the sample portfolio $\mathcal{P}^s$ in Table 4.8.

Table 4.8 Cash flow scheme for the sample portfolio $\mathcal{P}^s$

	$t = 0$	$t = 1$	$t = 2$	$t = 3$	$t = 4$	$t = 5$	$t = 6$
PREM_INC	30,035	24,416	18,511	11,458	5,098	711	0
EXPENSES	6,345	5,299	3,891	2,682	1,393	436	0
DTH_OUT		11,974	10,845	8,972	6,661	3,907	1,402
MAT_OUT		311,326	282,861	264,759	522,069	288,295	341,183
NET_CF	−23,690	304,183	279,086	264,955	525,025	291,927	342,585
PV^{mc}(NET_CF)	1,878,051						

Note that, for example, the death outgo and the maturity outgo for $t = 1$ occur at the end of the first year whereas the premium income and the expenses for $t = 1$ are to be taken as cash flows at the beginning of the second year. Hence the meaning of $t = j$ is twofold. For payments falling due in advance it denotes the beginning of year $j + 1$, for payments falling due in arrear it denotes the end of year j.

Remember the meaning of PV^{mc}(NET_CF). It is that amount we effectively need and simultaneously the initial value for the policy account P referring to the model's underlying balance sheet pictured in Table 3.1. From the data we further get the initial value for the assets $A_0 = 1{,}826{,}262$. The undistributed surplus is zero at start and so we have the values as shown in Table 4.9.

Table 4.9 Initial balance sheet entries for the sample portfolio $\mathcal{P}^s$

Quantity	Value
A_0	1,826,262
P_0	1,878,051
U_0	0
B_0	−51,789

These are all the policy data we need for modelling purposes. Besides we have to simulate the financial market in order to invest our assets A. As outlined in Sect. 3.7 we suggest to drawing up an asset portfolio based on the asset allocation defined in Table 4.10.

Table 4.10 A sample asset allocation

Asset	Weight
Equity fund	0.06
Property fund	0.14
1 year ZCB	0.05
3 year ZCB	0.02
5 year ZCB	0.07
10 year ZCB	0.20
15 year ZCB	0.10
20 year ZCB	0.36
Total	1.00

Using the stochastic scenario files for the equity, the property and the ZCB price process as presented in Chap. 3 the reference portfolio can be created.[3] For the current example only one simulation of the reference portfolio's price over the first 6 years corresponding to the portfolio duration is needed. See the sample price path in Table 4.11.

Table 4.11 One sample price path for the reference portfolio based on the asset allocation as defined in Table 4.10

Time	$t = 0$	$t = 1$	$t = 2$	$t = 3$	$t = 4$	$t = 5$	$t = 6$
Value (INDEX)	1.0000	1.0170	1.0343	1.1428	1.2446	1.1733	1.1900

Finally suppose the modelling parameters shown in Table 4.12.

Table 4.12 Sample modelling parameters

Parameter	Value
α (distribution ratio)	0.5
β (target buffer ratio)	0.05
sh_margin (shareholder margin)	0.01
i_u (bonus interest rate)	0.02
max_TB_rate (maximum terminal bonus relative to the maturity outgo)	0.4

After having outlined the essential assumptions and preconditions the simulation may be started by applying the algorithm presented in Sect. 4.2. Find the resulting values in Table 4.13.

Even though this example is entirely based on real existing data an extensive analysis would actually not make too much sense since only a small data basis has been taken and the input parameters were set arbitrarily. Analysis and interpretation of the modelling results is the main content of the empirical analysis in Chap. 5. However that example has a strong illustrative character and so I will conclude with the calculation of the option's values embedded in the sample portfolio $\mathcal{P}^s$ according to Formulae 4.11–4.13.

$$\text{Value of interest rate guarantee} = 24{,}130$$
$$\text{Value of dividend option} = 109{,}147$$
$$\text{Value of shareholder option} = 16{,}309$$

After having terminated that example all theoretical fundamentals are known to begin the empirical part of this thesis containing a detailed empirical analysis.

[3] Here ZCB prices were generated using the G2++ model.

Table 4.13 Results for the sample simulation run of the entire modelling approach

	$t = 0$	$t = 1$	$t = 2$	$t = 3$	$t = 4$	$t = 5$
Calculations at the beginning of a period (year)						
$A^{bp}(t)$	1,849,952	1,577,218	1,324,962	1,188,010	725,314	336,749
$P^{bp}(t)$	1,901,741	1,607,733	1,347,877	1,107,561	608,117	332,585
$U^{bp}(t)$	0	0	0	12,755	46,968	25,502
$B^{bp}(t)$	−51,789	−30,515	−22,915	67,694	70,228	−21,338
Calculations at the end of a period (year), before bonus calculations						
ACC_BONUS	0	0	0	6,049	22,406	26,012
TERM_BONUS	0	0	0	31,405	32,681	0
$A^{bb}(t+)$	1,558,101	1,310,342	1,190,222	727,653	336,474	−27,055
$P^{bb}(t+)$	1,588,616	1,333,257	1,098,785	604,412	332,310	0
$U^{bb}(t+)$	0	0	0	6,961	25,502	0
$B^{bb}(t+)$	−30,515	−22,915	91,437	116,280	−21,338	−27,055
Calculations at the end of a period (year), after bonus calculations						
SH_PROFIT	0	0	10,988	6,044	0	0
PH_PROFIT	0	0	12,755	40,008	0	0
$A^{ab}(t+)$	1,558,101	1,310,342	1,179,234	721,609	336,474	−27,055
$P^{ab}(t+)$	1,588,616	1,333,257	1,098,785	604,412	332,310	0
$U^{ab}(t+)$	0	0	12,755	46,968	25,502	0
$B^{ab}(t+)$	−30,515	−22,915	67,694	70,228	−21,338	−27,055

Chapter 5
An Empirical Analysis Using the Entire Modelling Approach

This chapter performs an extended empirical analysis of a large set of existing in-force insurance policies using the entire modelling approach as presented in the previous chapter.[1] The structure of this analysis is supposed to be as follows: First of all, I will carry out the so called *basic simulation run* assuming consequently realistic or, to my mind, the most appropriate parameter specifications. By doing this we will get a first impression of the market consistent values of the interest rate guarantee, the dividend and the shareholder option. These values will be considered under three sets of different surplus parameters α, the distribution ratio, and β, the target buffer ratio. Therefore, recall the calculation of the policyholder profit as presented in the previous chapter:

$$\text{PH\_PROFIT}(t^+) = \max\{0, \alpha \cdot (B^{bb}(t^+) - \text{SH\_PROFIT}(t^+) - \beta P^{bb}(t^+))\}$$

Based on this basic simulation run, I will analyse the impacts on the values of the embedded options by varying certain components of the model. This sensitivity analysis incorporates:

- A variation of the shareholder margin,
- A variation of the interest rate model to generate stochastic ZCB prices,
- A variation of the asset allocation,
- An analysis of as at the end of 2004 data,
- A variation of the market interest rates,
- A variation of the surplus distribution mechanism and the surplus determination and
- A shock of the mortality rates.

[1] Confer Table 4.1 for detailed information on the policy data.

N. Rüfenacht, *Implicit Embedded Options in Life Insurance Contracts*, Contributions to Management Science, DOI 10.1007/978-3-7908-2843-6_5,
© Springer-Verlag Berlin Heidelberg 2012

These shall be the key figures in order to approximate the targeted values as outlined in Table 4.8. Note that an exact determination of these values is not necessary. Its calculation would be expensive and we will get the relevant insights on the behaviour of the embedded options by performing the sensitivity analysis already. And this is finally the essential.

5.1 The Basic Simulation Run

The idea of carrying out the basic simulation run is to define a reference simulation run as at the end of 2009 using consequently realistic parameter specifications. The following sensitivity analysis will always refer to this basic simulation run. In case of the ZCB price process it is not that obvious which interest rate model should be selected a priori. However, in my opinion, the most appropriate for the purpose of this modelling approach is the G2++ model. Unlike the CIR model, the G2++ model matches the current term structure of interest rates and measured in terms of the mean squared error term Ψ, as defined by Eq. 3.55, it features a better calibration fit than the HW model. Find a description of what is meant by best estimate assumptions for the basic simulation run below.

Component	Description
Share price process	A stochastic scenario file containing 5,000 simulation paths for the share prices was generated based on the parameters estimated in Sect. 3.3.1. The share price process is supposed to map the SPI and to be correlated to the property price process
Property price process	A stochastic scenario file containing 5,000 simulation paths for the property prices was generated based on the parameters estimated in Sect. 3.3.1. The property price process is supposed to map the IAZI index and to be correlated to the share price process
ZCB price process	A stochastic scenario file containing 5,000 simulation paths for the prices for ZCBs with maturities of 1, 3, 5, 10, 15 and 20 years was generated using the G2++ interest rate model based on the parameters estimated in Sect. 3.5.3

(continued)

Component	Description
Reference portfolio	The reference portfolio was generated according to the considerations of Sect. 3.7 using the stochastic scenario files for the share price process, the property price process and the ZCB price process as outlined above. The underlying asset allocation will be defined in the subsequent
Asset allocation	Table 3.2 shows the summarised life insurance asset portfolio of Helvetia as at the end of 2009. In order to define a realistic asset allocation, the weights for shares (4.5%) and property (16.0%) shall be taken directly from this table. A deeper analysis of the Helvetia's bond portfolio shows that 2.5% of the bonds have an origin maturity shorter or equal to 3.5 years, 11.6% have an origin maturity longer than 3.5 years but shorter or equal to 7.5 years, 34.7% have an origin maturity longer than 7.5 years but shorter or equal to 12.5 years, 26.4% have an origin maturity longer than 12.5 years but shorter or equal to 17.5 years, and finally 24.8% have an origin maturity longer than 17.5 years. So the overall weight of 55% of bonds shall be further subdivided in classes of 3, 5, 10, 15 and 20 year ZCBs[2]
	I will further assume cash (4.0%) and mortgages (14.6%) to be modelled as a 1 year ZCB and a 20 year ZCB respectively. Neglecting the category "Else" (5.9%) there is 94.1% of the total asset amount to be allocated forming a feasible asset allocation. However, I will define the realistic asset allocation by the following weights w_i
	$w_{\text{shares}} = 0.045/0.941 = 0.0478$ $w_{\text{property}} = 0.16/0.941 = 0.1700$ $w_{1\,\text{y ZCB}} = 0.04/0.941 = 0.0425$ $w_{3\,\text{y ZCB}} = 0.025 \cdot 0.55/0.941 = 0.0146$ $w_{5\,\text{y ZCB}} = 0.116 \cdot 0.55/0.941 = 0.0678$ $w_{10\,\text{y ZCB}} = 0.347 \cdot 0.55/0.941 = 0.2028$

(continued)

[2]The raw data are not publicly available.

Component	Description
	$w_{15\,\text{y ZCB}} = 0.264 \cdot 0.55/0.941 = 0.1543$ $w_{20\,\text{y ZCB}} = (0.248 \cdot 0.55 + 0.146)/0.941 = 0.3001$
Mortality rates	To determine the net cash flow NET_CF the mortality rates from table EKM/F_01_05_II are used[3]
Discount rates	For discounting purposes the current swap rate curve as at 31 December 2009 is used
Policy data	All policy data are taken as at the end of 2009. For further information on the policy data confer Table 4.1
Bonus	As mentioned in the previous section the algorithm is based on a portfolio level and not on an individual policy level. Thus it is necessary either to assume premium reduction or interest-bearing accumulation as the bonus distribution mechanism. For the purpose of the basic simulation run, interest-bearing accumulation is assumed since this is the case for the majority of the policies
sh_margin	The shareholder margin is supposed to be 0.005, that is considered as a reasonable value in practice
i_u	The surplus accumulation rate i_u is assumed to be equal to the current 10 year swap rate $i_u = s_{10} = 0.025$
max_TB_rate	The maximum yearly terminal bonus relative to the maturity outgo is supposed to be 0.25
Number of simulations	All simulation runs are based on a number of 5,000

Now all the necessary quantities are defined to perform the basic simulation run. It will be self-evident to maintain the structure that categorises the data into its types of insurance: Endowment, whole life, term life and pure endowment insurance as suggested in Table 4.1. A further subdivision will be made for single premiums and regular premiums.

[3]Confer page 90 for the definition of the NET_CF.

The values of the different options will be evaluated on a category level only. Since each of the four categories consists of four to twelve portfolios, an analysis on an individual portfolio level would produce a vast quantity of data and confuse the reader. However, there is a brief outline on the initial values of the balance sheet entries A, P and B including the technical interest rate $i_{\text{tech.}}$ for each of the four categories shown in Tables 5.1–5.4. Considering the results for single premiums in Tables 5.1 and 5.4, we recognise that the initial values of the asset portfolio A_0 are usually smaller than the initial values of the policy account P_0 and hence the buffers at start B_0 are negative for endowment insurance and pure endowment insurance. In other words, today the amount we effectively have (A_0) is smaller than the amount we effectively need (P_0).

Remember Formula 4.8. P_0 is defined as the present value of a portfolio's net cash flow discounted with the current swap rates. On the other hand, A_0 is defined as the statutory policy reserve that is determined similarly, but using the technical interest rate $i_{\text{tech.}}$ as the discount rate. And what we see is that the higher $i_{\text{tech.}}$ is, the lower the initial value of the buffer B_0 is. So the difference between A_0 and P_0 basically reflects the different valuation principles, statutory or market consistent.

As shown in Table 5.3, the initial values for the policy account P_0 are predominantly negative for term life insurance. At first sight, this result surprises. But realise that for this type of insurance there is no survival benefit at the end and so the structure of the benefit scheme is significantly different compared the to endowment insurance. So bearing in mind that P_0 is a present value discounted with the current swap rate curve, this negative values can be verified: The longer a term life insurance runs, the higher is the expected outflow for the mortality benefit. Payments in the near future are discounted with a lower interest rate than payments in the far future and this asymmetry finally leads to the negative initial values of the policy account P_0.

As mentioned at the beginning of the current chapter, we will consider the values of the embedded options for three sets of surplus parameter combinations α and β. Because of this, each interpretation of the results will be based on the values shown, that means I will consequently refer to the following sets of parameter combinations only.

There is no defined rule how to chose these parameters but in my opinion these are just reasonable combinations that offer interesting aspects for the purpose of this thesis. Thus, it makes sense to keep them constant throughout the whole chapter. The parameter combinations shall be as follows.

$$\alpha = 0.8, \ \beta = 0.1$$

$$\alpha = 0.4, \ \beta = 0.1$$

$$\alpha = 0.4, \ \beta = 0.05$$

Note that the values of these parameters are not directly comparable with the ones estimated in Grosen and Jørgensen (2002b) since their surplus determination structurally differs from the one that is applied here.

Table 5.1 Initial values of the balance sheet entries A, P and B for endowment insurance. Amounts of money are denoted in Swiss Francs

TG	No. pols	$i_{\text{tech.}}$	A_0	P_0	B_0
Single premium					
1970	369	0.0325	4,570,185	4,707,196	−137,011
1980	5,824	0.0300	159,180,183	159,961,832	−781,649
1994	5,291	0.0350	145,160,091	152,918,220	−7,758,130
1996	2,143	0.0350	114,014,483	122,086,401	−8,071,918
1999	5,080	0.0250	427,735,330	434,589,313	−6,853,983
2003	5,405	0.0200	231,834,204	231,730,742	103,462
Total	24,112		1,082,494,476	1,105,993,704	−23,499,229
Regular premium					
1970	2,318	0.0325	123,565,286	125,250,510	−1,685,225
1980	9,653	0.0300	379,599,790	362,108,856	17,490,934
1994	21,867	0.0350	496,718,988	497,844,701	−1,125,712
1996	7,880	0.0350	232,796,121	242,411,782	−9,615,661
1999	9,077	0.0250	194,345,208	152,917,703	41,427,505
2003	21,451	0.0200	131,924,696	−13,187,316	145,112,012
Total	72,246		1,558,950,089	1,367,346,236	191,603,853

Table 5.2 Initial values of the balance sheet entries A, P and B for whole life insurance. Amounts of money are denoted in Swiss Francs

TG	No. pols	$i_{\text{tech.}}$	A_0	P_0	B_0
Single premium					
1980	724	0.0300	42,587,869	37,641,562	4,946,307
1994	92	0.0350	2,716,070	2,605,084	110,986
1996	35	0.0350	594,581	627,576	−32,995
1999	42	0.0250	1,156,378	1,081,782	74,596
2003	152	0.0200	6,586,457	5,745,532	840,925
Total	1,045		53,641,355	47,701,536	5,939,819
Regular premium					
1980	749	0.0300	8,495,508	7,971,111	524,397
1994	93	0.0350	938,547	922,378	16,168
1996	14	0.0350	501,191	526,766	−25,575
1999	19	0.0250	281,757	248,706	33,051
2003	55	0.0200	1,563,731	832,159	731,572
Total	930		11,780,734	10,501,120	1,279,613

Before presenting the results of the basic simulation run, recall how the values $V(\cdot)$ of the considered options are defined in Sect. 4.3.

$$V(\text{interest rate guarantee}) := -\mathbb{E}[B^{ab}(T^+) \cdot (1 + s_{T+1})^{-(T+1)} \mid B(T^+) < 0],$$

Table 5.3 Initial values of the balance sheet entries A, P and B for term life insurance. Amounts of money are denoted in Swiss Francs

TG	No. pols	$i_{\text{tech.}}$	A_0	P_0	B_0
Regular premium					
1980	2,510	0.0300	14,648,230	1,938,376	12,709,854
1994	1,723	0.0350	7,815,816	−1,555,442	9,371,258
1996	1,249	0.0350	3,905,389	−502,020	4,407,410
1999	2,142	0.0250	4,518,666	−3,263,946	7,782,612
2003	2,024	0.0200	2,855,437	−5,318,327	8,173,765
2006	1,533	0.0200	699,300	−5,939,297	6,638,597
Total	11,181		34,442,838	−14,640,656	49,083,496

Table 5.4 Initial values of the balance sheet entries A, P and B for pure endowment insurance. Amounts of money are denoted in Swiss Francs

TG	No. pols	$i_{\text{tech.}}$	A_0	P_0	B_0
Single premium					
1970	231	0.0300	8,549,347	8,849,997	−300,651
1980	319	0.0350	12,979,086	14,005,275	−1,026,190
Total	550		21,528,433	22,855,272	−1,326,841
Regular premium					
1970	154	0.0300	6,569,139	7,000,573	−431,433
1980	1,219	0.0350	31,136,888	34,254,285	−3,117,396
Total	1,373		37,706,027	41,254,858	−3,548,829

$$V(\text{dividend option}) := \mathbb{E}\Big[\sum_{t=0}^{T}\Big(\text{TERM\_BONUS}(t^+) + \text{ACC\_BONUS}(t^+) + \text{PREM\_RED}(t+1)\Big)\cdot(1+s_{t+1})^{-(t+1)}\Big]$$

and

$$V(\text{shareholder option}) := \mathbb{E}\Big[\sum_{t=0}^{T}(\text{SH\_PROFIT}(t^+))\cdot(1+s_{t+1})^{-(t+1)}\Big],$$

with s_t the current $t-$year swap rate. Throughout this chapter, the expected values $\mathbb{E}[\cdot]$ are estimated by the mean values over all of the 5,000 simulation runs. Find the results for the value of the interest rate guarantee, the dividend and the shareholder option in Tables 5.6, 5.7 and 5.8 respectively.

What do these results tell us? From the perspective of the insurer the values of these embedded options quantify the market consistent liability as at the end of 2009 under best estimate assumptions. As stated in Sect. 4.3 it is essential to understand the exact meaning of the values of these options. Since the dividend and the shareholder option are implicitly hedged by definition already, the insurer does not need to set up a reserve for these options today. Its values basically quantify the worth additionally offered to the policyholders and the shareholders due to the currently applied surplus policy.

Different than surpluses the insurer guarantees crediting a fixed minimum interest rate on the savings premiums and this guarantee is neither implicitly nor explicitly hedged in this model. Thus, the value of the interest rate guarantee actually quantifies the amount the insurer should reserve today in order to meet this future obligations.

From the perspective of the policyholder and the shareholder, the values quantify the market consistent value of their additional but non guaranteed benefits as at the end of 2009.

Indeed, the absolute values of, for example, the interest rate guarantee as shown in Table 5.6 are a first interesting result, but they are hardly interpretable without any relation. Therefore, its "true" values are expressed by the absolute option values relative to the initial assets A_0. We constitute that the total values of the interest rate guarantee are mainly dominated by the endowment insurance. This should not further surprise since endowment insurance hold the majority of all data under consideration.

For portfolios paid for regular premiums the relative values of the interest rate guarantee are lower than for single premium policies. As we will see in the subsequent, this difference becomes even more obvious for the dividend option, and so I will return to this charge below when discussing the latter.

What influence do the distribution ratio α and the target buffer ratio β have on the interest rate guarantee? We observe that the more we distribute to the policyholder (increasing α), the more we take a loss (increase of the values of the interest rate guarantee). On the other hand, increasing the target buffer ratio β delays the allocation of possible surpluses and thus decreases the values of the interest rate guarantee. Depending on the type of insurance, the variation of α and β has a stronger or weaker impact on the resulting values. The impact tends to be stronger for insurance types that do not have a guaranteed survival benefit, like for term life and whole life insurance.

These observations do not further surprise since they could have been expected a priori and, moreover, the current results coincide with the default probabilities in Grosen and Jørgensen (2002a), Chap. 4.

Let us proceed and have a look at the values of the dividend option that are shown in Table 5.7.

The total values of the dividend option behave similar to the interest rate guarantee when varying parameters α and β. A higher distribution ratio increases the values of the dividend option, a higher target buffer ratio decreases its values.

At a second sight, we remark that the value of the dividend option for term life insurance and whole life insurance is greater in case of the lower distribution ratio $\alpha = 0.4$ and not, as probably expected, the other way around. This might appear a little curious, but obviously it is, from the perspective of the policyholders, a more valuable strategy when surpluses are allocated more conservatively. If the insurer sets its α high, the buffer B sinks quite fast and might even fall below its targeted value $\beta \cdot P^{bb}$, whereas for a lower α it still would be possible to allocate surpluses in future years. Additionally, the asset portfolio A and consequently the returns on it are the lower, the more surpluses are paid out. This is how that curiosity can be explained.

Comparing the results for single and regular premiums, we see that there is a huge difference for the relative values of the dividend option, too. I have mentioned that matter already when discussing the interest rate guarantee. There are mainly two reasons which explain this difference.

The relative values are obtained by dividing the absolute values by the initial assets A_0, the sum of current policy and surplus reserves. Naturally the average amount of A_0 per policy is remarkably greater for single premium policies. Hence it is difficult to directly contrast theses ratios with each other.

Second, remember that we are working with in-force policies and consequently there is no future premium income for single premium policies. Each single premium has been paid in the past when the respective policies were issued and so we have the premiums included in the statutory policy reserves. And since these reserves are statutory, they are not market consistent. Unlike the single premium case, policies paid for regular premiums still earn premiums during the simulation run, but now the premium income is valued market consistently. This leads to a different valuation result of the policy reserves.

So we see that given these explanations the results for single and regular premium policies must be regarded each for itself. Finally we shall consider the values of the shareholder option. As shown in Table 5.8, the values change contrary to the values of the dividend option.

The behaviour of the values of this option is plausible in most of the cases. Even though the shareholder profit is calculated prior to the policyholder profit, its value indirectly depends on the surplus distribution parameters α and β. The more we distribute to the policyholders (increasing α or decreasing β), the less remains in the future for the shareholders and vice versa. And again, we observe higher relative option values for policies paid for regular premiums than for policies paid for single premiums.

The only abnormal values appear for term life insurance where the value of the shareholder option seems to be very small. Remember the initial value for the policy account P_0, that is clearly below zero as shown in Table 5.3. As long as $P(t)$ is below zero, no shareholder profit will be allocated according to the current model definition. On the other hand, a policyholder profit is feasible, even for a negative $P(t)$.[4]

Another interesting result appears when calculating the option values per policy. We constitute that for the dividend and the shareholder option, the values per policy are higher for regular premiums, whereas it is the opposite way around for the interest rate guarantee. Using these results one can argue that this model

[4]See the *calculations at the end of a period, after bonus calculations (ab)* on page 95 for the respective formulae.

identifies regular premium life insurance as the more profitable ones. Therefor, confer Table 5.5 below that summarises these values based on all the available data under best estimate assumptions and for a representative set of surplus parameters, that is to say a distribution ratio $\alpha = 0.8$ and a target buffer ratio $\beta = 0.1$.

Table 5.5 Comparison of the per policy values in Swiss Francs of the interest rate guarantee, the dividend and the shareholder option under best estimate assumptions. The underlying data basis consists of 25,707 single premium policies and 85,730 regular premium policies

$\alpha = 0.8, \beta = 0.10$	Single premium	Regular premium
Total interest rate guarantee	39,497,246	42,686,809
Interest rate guarantee per policy	1,536	498
Total dividend option	89,501,240	893,824,435
Dividend option per policy	3,482	10,426
Total shareholder option	18,358,262	102,467,939
Shareholder option per policy	714	1,195

Since the basic simulation run defines the reference simulation run for the following sensitivity analysis, it makes sense to provide a brief summary of it in the subsequent section.

Interest rate guarantee/basic simulation run

Table 5.6 Absolute values of the interest rate guarantee in Swiss Francs for the basic simulation run. The values in brackets denote the absolute option value relative to the initial assets A_0

Category	Surplus parameter combination		
Single premium	$\alpha = 0.8, \beta = 0.10$	$\alpha = 0.4, \beta = 0.10$	$\alpha = 0.4, \beta = 0.05$
Endowment	37,462,100 (0.0346)	34,962,729 (0.0323)	37,323,594 (0.0345)
Whole life	640,955 (0.0119)	326,398 (0.0061)	568,315 (0.0106)
Pure endowment	1,394,191 (0.0648)	1,364,172 (0.0634)	1,396,683 (0.0649)
Total	39,497,246 (0.0341)	36,653,299 (0.0317)	39,288,592 (0.0339)
Regular premium	$\alpha = 0.8, \beta = 0.10$	$\alpha = 0.4, \beta = 0.10$	$\alpha = 0.4, \beta = 0.05$
Endowment	41,314,171 (0.0265)	32,720,607 (0.0210)	42,522,974 (0.0273)
Whole life	222,650 (0.0189)	155,177 (0.0132)	224,926 (0.0191)
Term life	561,059 (0.0163)	224,307 (0.0065)	238,767 (0.0069)
Pure endowment	588,929 (0.0156)	498,084 (0.0132)	646,830 (0.0172)
Total	42,686,809 (0.0260)	33,598,175 (0.0205)	43,633,497 (0.0266)

Dividend option/basic simulation run

Table 5.7 Absolute values of the dividend option in Swiss Francs for the basic simulation run. The values in brackets denote the absolute option value relative to the initial assets A_0

Category	Surplus parameter combination		
Single premium	$\alpha = 0.8, \beta = 0.10$	$\alpha = 0.4, \beta = 0.10$	$\alpha = 0.4, \beta = 0.05$
Endowment	72,789,688 (0.0672)	70,702,606 (0.0653)	72,425,757 (0.0669)
Whole life	15,961,518 (0.2976)	16,281,035 (0.3035)	16,230,906 (0.3026)
Pure endowment	750,034 (0.0348)	713,023 (0.0331)	748,280 (0.0348)
Total	89,501,240 (0.0773)	87,696,664 (0.0758)	89,404,943 (0.0772)
Regular premium	$\alpha = 0.8, \beta = 0.10$	$\alpha = 0.4, \beta = 0.10$	$\alpha = 0.4, \beta = 0.05$
Endowment	825,006,421 (0.5292)	808,561,657 (0.5187)	830,828,880 (0.5329)
Whole life	6,209,928 (0.5271)	6,475,320 (0.5497)	6,449,594 (0.5475)
Term life	58,754,653 (1.7059)	60,111,998 (1.7453)	60,144,617 (1.7462)
Pure endowment	3,853,433 (0.1022)	3,693,505 (0.0980)	3,970,558 (0.1053)
Total	893,824,435 (0.5441)	878,842,480 (0.5349)	901,393,649 (0.5487)

Shareholder option/basic simulation run

Table 5.8 Absolute values of the shareholder option in Swiss Francs for the basic simulation run. The values in brackets denote the absolute option value relative to the initial assets A_0

Category	Surplus parameter combination		
Single premium	$\alpha = 0.8, \beta = 0.10$	$\alpha = 0.4, \beta = 0.10$	$\alpha = 0.4, \beta = 0.05$
Endowment	16,045,581 (0.0148)	17,076,659 (0.0158)	15,729,586 (0.0145)
Whole life	2,126,324 (0.0396)	2,333,176 (0.0435)	2,137,242 (0.0398)
Pure endowment	186,357 (0.0087)	197,282 (0.0092)	182,066 (0.0085)
Total	18,358,262 (0.0159)	19,607,117 (0.0169)	18,048,894 (0.0156)
Regular premium	$\alpha = 0.8, \beta = 0.10$	$\alpha = 0.4, \beta = 0.10$	$\alpha = 0.4, \beta = 0.05$
Endowment	100,546,363 (0.0645)	109,608,056 (0.0703)	99,933,922 (0.0641)
Whole life	739,546 (0.0628)	809,123 (0.0687)	735,201 (0.0624)
Term life	145,734 (0.0042)	210,446 (0.0061)	207,321 (0.0060)
Pure endowment	1,036,296 (0.0275)	1,079,708 (0.0286)	984,099 (0.0261)
Total	102,467,939 (0.0624)	111,707,333 (0.0680)	101,860,543 (0.0620)

5.1.1 *A Summary of the Key Results of the Basic Simulation Run*

The basic simulation run gives a first impression of the as at the end of 2009 market consistent values of the interest rate guarantee, the dividend and the shareholder option, given throughout realistic modelling assumptions. To further support the informative value of the results and to consolidate the modelling approach itself, a sensitivity analysis is outstanding.

First of all, the initial values for the balance sheet entries A_0, the asset portfolio, P_0, the policy account, and B_0, the buffer, are presented. Find the total initial values of the balance sheet entries summarised in Table 5.9.

Table 5.9 The total initial values of the balance sheet entries across all categories of insurance in the basic simulation run. The values are denoted in Swiss Francs

	Single premium	Regular premium
A_0	1,157,664,264	1,642,879,688
P_0	1,176,550,512	1,404,461,558
B_0	−18,886,248	238,418,130
Number of policies	25,707	85,730

The different embedded options are considered for a set of endowment, pure endowment, term life and whole life insurance, separated by the premium payment type. The values are determined under three sets of different surplus parameters α, the distribution ratio, and β, the target buffer ratio. Table 5.10 summarises these values based on all the available data under best estimate assumptions and for a representative set of surplus parameters, that is to say $\alpha = 0.8$ and $\beta = 0.1$. Besides, the shareholder margin sh_margin is equal to 0.005. The relative values of the embedded options are obtained by dividing by the current policy reserves including the current surplus reserves, that is equal to the initial value of the asset portfolio A_0.

What do these results tell us? From the perspective of the insurer the values of these embedded options quantify the market consistent liability as at the end of 2009 under best estimate assumptions. Since the dividend and the shareholder option are implicitly hedged by definition already, the insurer does not need to set up a reserve for these options today. Their values basically quantify the worth additionally offered to the policyholders and the shareholders due to the currently applied surplus policy.

Different than surpluses the insurer guarantees crediting a fixed minimum interest rate on the savings premiums and this guarantee is neither implicitly nor explicitly hedged in this model. Thus, the value of the interest rate guarantee actually quantifies the amount the insurer should reserve today in order to meet this future obligations.

Table 5.10 Total values in Swiss Francs of the interest rate guarantee, the dividend and the shareholder option under best estimate assumptions. The underlying data basis consists of 25,707 single premium policies and 85,730 regular premium policies with an initial asset portfolio A_0 of 1,157,664,264 Swiss Francs (single premium) and 1,642,879,688 Swiss Francs (regular premium)

$\alpha = 0.8, \beta = 0.10$, sh_margin $= 0.005$	Single premium	Regular premium
Total interest rate guarantee	39,497,246	42,686,809
Interest rate guarantee per policy	1,536	498
Interest rate guarantee relative to A_0	0.0341	0.0260
Total dividend option	89,501,240	893,824,435
Dividend option per policy	3,482	10,426
Dividend option relative to A_0	0.0773	0.5441
Total shareholder option	18,358,262	102,467,939
Shareholder option per policy	714	1,195
Shareholder option relative to A_0	0.0159	0.0624
Total of all options	147,356,748	1,038,979,183
Total of all options per policy	5,732	12,119
Total of all options relative to A_0	0.1273	0.6324

From the perspective of the policyholder and the shareholder, the values quantify the market consistent value of their additional but non guaranteed benefits as at the end of 2009.

The interested reader might find additional information on the characteristics of this basic simulation run in Appendix B, where the coefficient of variation for the interest rate guarantee and the dividend option, the values of the terminal bonuses and the default probabilities for the basic simulation run are presented.

I will proceed with the sensitivity analysis of the basic simulation run in the subsequent section. Its idea is to analyse the different values for the interest rate guarantee, the dividend and the shareholder option when varying certain components of the models. The components to be varied are outlined at the beginning of the current chapter. In the following the shareholder margin shall be the first parameter to be varied.

5.2 Variation of the Shareholder Margin

After having performed a basic simulation run to determine the different option values under best estimate assumptions, it is the moment to deeper analyse the modelling approach by an extensive sensitivity analysis. The procedure is based on the end of 2009 basic simulation run as presented in Sect. 5.1 whilst one component of it is varied. This sensitivity analysis starts with the variation of the shareholder margin, that mainly determines the shareholder profit. Therefore, recall the calculation of the shareholder profit as presented in the previous chapter:

$$\text{SH\_PROFIT}(t^+) = \max\{0, \min\{B^{bb}(t^+), \text{sh\_margin} \cdot P^{bb}(t^+)\}\}$$

The best estimate value of sh_margin is fixed at 0.005, the scope of variation shall be given by the set $S = \{0.001, 0.003, 0.007, 0.009\}$.

For the remainder of this sensitivity analysis we will solely consider the total value of a particular option across all categories of insurance types, separated by the premium payment type. This means that the option values are calculated separately for each category and then added. Again, a deeper segregation would complicate a clear and comprehensible interpretation of the results.

The results for the current sensitivity analysis are shown below. Tables 5.11, 5.12 and 5.13 present the values under various shareholder margins for the interest rate guarantee, the dividend and the shareholder option respectively. The reference values from the basic simulation run are denoted in bold letters.

Interest rate guarantee/variation of the shareholder margin

Table 5.11 Values of the interest rate guarantee in Swiss Francs using different shareholder margins. The values in brackets denote the relative difference to the basic simulation run

Component	Surplus parameter combination		
Single premium	$\alpha = 0.8, \beta = 0.10$	$\alpha = 0.4, \beta = 0.10$	$\alpha = 0.4, \beta = 0.05$
sh_margin = 0.001	38,589,706	35,775,755	38,160,913
	(−0.0230)	(−0.0239)	(−0.0287)
sh_margin = 0.003	39,025,165	36,194,015	38,703,044
	(−0.0120)	(−0.0125)	(−0.0149)
sh_margin = 0.005	**39,497,246**	**36,653,299**	**39,288,592**
	(0.0000)	(0.0000)	(0.0000)
sh_margin = 0.007	39,998,072	37,147,230	39,910,234
	(0.0127)	(0.0135)	(0.0158)
sh_margin = 0.009	40,528,881	37,676,619	40,567,419
	(0.0261)	(0.0279)	(0.0325)
Regular premium	$\alpha = 0.8, \beta = 0.10$	$\alpha = 0.4, \beta = 0.10$	$\alpha = 0.4, \beta = 0.05$
sh_margin = 0.001	40,819,344	30,980,146	41,375,682
	(−0.0437)	(−0.0779)	(−0.0517)
sh_margin = 0.003	41,723,687	32,247,955	42,476,868
	(−0.0226)	(−0.0402)	(−0.0265)
sh_margin = 0.005	**42,686,809**	**33,598,175**	**43,633,497**
	(0.0000)	(0.0000)	(0.0000)
sh_margin = 0.007	43,726,457	35,029,715	44,823,467
	(0.0244)	(0.0426)	(0.0273)
sh_margin = 0.009	44,844,568	36,538,828	46,059,441
	(0.0505)	(0.0875)	(0.0556)

Dividend option/variation of the shareholder margin

Table 5.12 Values of the dividend option in Swiss Francs using different shareholder margins. The values in brackets denote the relative difference to the basic simulation run

Component	Surplus parameter combination		
Single premium	$\alpha = 0.8, \beta = 0.10$	$\alpha = 0.4, \beta = 0.10$	$\alpha = 0.4, \beta = 0.05$
sh_margin = 0.001	105,662,500	105,317,709	105,167,330
	(0.1806)	(0.2009)	(0.1763)
sh_margin = 0.003	97,396,864	96,319,532	97,074,491
	(0.0882)	(0.0983)	(0.0858)
sh_margin = 0.005	**89,501,240**	**87,696,664**	**89,404,943**
	(0.0000)	(0.0000)	(0.0000)
sh_margin = 0.007	81,980,230	79,476,720	82,175,280
	(−0.0840)	(−0.0937)	(−0.0809)
sh_margin = 0.009	74,862,275	71,692,064	75,395,287
	(−0.1636)	(−0.1825)	(−0.1567)
Regular premium	$\alpha = 0.8, \beta = 0.10$	$\alpha = 0.4, \beta = 0.10$	$\alpha = 0.4, \beta = 0.05$
sh_margin = 0.001	993,649,468	987,627,245	1,001,364,116
	(0.1117)	(0.1238)	(0.1109)
sh_margin = 0.003	942,662,854	932,069,765	950,264,833
	(0.0546)	(0.0606)	(0.0542)
sh_margin = 0.005	**893,824,435**	**878,842,480**	**901,393,649**
	(0.0000)	(0.0000)	(0.0000)
sh_margin = 0.007	847,291,078	828,221,848	854,882,612
	(−0.0521)	(−0.0576)	(−0.0516)
sh_margin = 0.009	803,119,844	780,361,645	810,738,211
	(−0.1015)	(−0.1121)	(−0.1006)

Considering each of the options separately, we observe the following. First, the values of the interest rate guarantee as shown on Table 5.11 raise as the shareholder margin raises. Since there is no buffer element for this option as for the policyholder profit, the insurer bears an extra risk when granting a higher shareholder profit. Second, as apparent from Table 5.12, the more the shareholder gets, the less remains for the policyholder and thus the values of the dividend option behave in opposite direction to the values of the shareholder option. There we make the third observation. In Table 5.13 we can see that there is approximately a linear relation between the shareholder margin and the values of the shareholder option.

Let us now compare the aggregated values of dividend and shareholder option with the values of the interest rate guarantee. Suppose, for example, a fixed parameter combination $\alpha = 0.8, \beta = 0.1$. As mentioned already, the interest

Shareholder option/variation of the shareholder margin

Table 5.13 Values of the shareholder option in Swiss Francs using different shareholder margins. The values in brackets denote the relative difference to the basic simulation run

Component	Surplus parameter combination		
Single premium	$\alpha = 0.8, \beta = 0.10$	$\alpha = 0.4, \beta = 0.10$	$\alpha = 0.4, \beta = 0.05$
sh_margin = 0.001	3,813,922	4,061,859	3,774,825
	(−0.7923)	(−0.7928)	(−0.7909)
sh_margin = 0.003	11,233,413	11,979,930	11,079,255
	(−0.3881)	(−0.3890)	(−0.3862)
sh_margin = 0.005	**18,358,262**	**19,607,117**	**18,048,894**
	(0.0000)	(0.0000)	(0.0000)
sh_margin = 0.007	25,178,647	26,915,995	24,665,445
	(0.3715)	(0.3728)	(0.3666)
sh_margin = 0.009	31,671,974	33,879,770	30,919,027
	(0.7252)	(0.7279)	(0.7131)
Regular premium	$\alpha = 0.8, \beta = 0.10$	$\alpha = 0.4, \beta = 0.10$	$\alpha = 0.4, \beta = 0.05$
sh_margin = 0.001	21,292,118	23,212,805	21,227,134
	(−0.7922)	(−0.7922)	(−0.7916)
sh_margin = 0.003	62,718,836	68,368,106	62,427,201
	(−0.3879)	(−0.3880)	(−0.3871)
sh_margin = 0.005	**102,467,939**	**111,707,333**	**101,860,543**
	(0.0000)	(0.0000)	(0.0000)
sh_margin = 0.007	140,417,758	152,992,896	139,402,291
	(0.3704)	(0.3696)	(0.3686)
sh_margin = 0.009	176,495,758	192,074,159	175,041,385
	(0.7224)	(0.7194)	(0.7184)

rate guarantee increases as the shareholder margin increases. On the other hand, the aggregated value of the dividend and the shareholder option decreases as we raise the shareholder margin. Hence, by reducing the shareholder margin the insurer reduces its risk (lowering the value of the interest rate guarantee) and lifts the present value of the total profits he will pay out. But notice that there is a rub in it. In practice, the insurer cannot reduce the shareholder margin in any order since the shareholders' interests are an important part of the overall strategic considerations of any company.

In the upcoming section the G2++ interest rate model, that is used to generate the stochastic ZCB prices, will be replaced by the two other models introduced in Chap. 3: The CIR and the HW model.

5.3 Variation of the Interest Rate Model to Generate Stochastic ZCB Prices

In Chap. 3, three different interest rate models are presented, the CIR, the HW and the G2++ model. Calibrated to market instruments, one can use them to generate stochastic ZCB prices and, according to this, generate a stochastic scenario file that is needed for the modelling.

At the beginning of the current chapter I have explained why to use the G2++ model to generate the stochastic scenario files for the purpose of the basic simulation run. Unlike the CIR model, the G2++ model matches the current term structure of interest rates and measured in terms of the mean squared error Ψ, as defined by Eq. 3.55, it features a better calibration fit than the HW model.

However, remember that ZCBs are extremely important instruments with respect to the asset modelling process and therefore it is indispensable to consider the corresponding model risks. Thus, the CIR and the HW model shall be analysed, too.

Before considering the resulting option values, it is helpful to further compare each of the stochastic scenario files generated by the different interest rate models. In Sect. 3.6 we see how the mean annualised return on the ZCB prices evolves over time across the different models.[5] From these figures we can moderately anticipate what impacts the use of each model could have. But be aware of overvaluing these mean values, they do not give us any information on the variability of the ZCB prices. Therefore, see Fig. 5.1 that plots the mean standard deviation on 10 year ZCB prices for the CIR, the HW and the G2++ model over time, based on 5,000 simulation runs. This figure clearly shows that the ZCB prices generated according to the CIR model possess a remarkably lower variability than the ZCB prices generated according to the other models.

With this extra information we now have the necessary background to understand the current results. As we see in Figs. 3.15 and 5.1, the mean annualised return on the ZCB prices in the CIR model is comparable to the returns in the other models, but its mean standard deviation is apparently lower. Thus, using the CIR model reduces the volatility within the asset portfolio and, given the return on the ZCBs is sufficiently high, lowers the values of the interest rate guarantee. Observing Table 5.14, this is exactly what happens in the simulation run, the interest rate guarantee almost disappears in the case of using the CIR model. This result surprises and one consequence from this is that the CIR model is unsuitable for the purpose of this modelling approach as it significantly underestimates the interest rate risk.

Above the expression "sufficiently high" is used in conjunction with the return on the ZCBs without any further explanations. In plain language and neglecting the other investment instruments like shares and property, this means that the return on the ZCBs has to be above the technical interest rate, which is the case here

[5]Confer Figs. 3.15, 3.16 and 3.17.

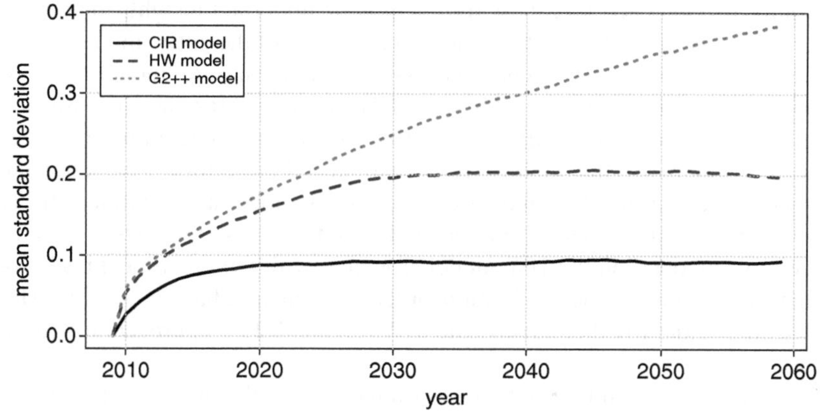

Fig. 5.1 Mean standard deviation on 10 year ZCB prices for the CIR, the HW and the G2++ model over time, based on 5,000 simulation runs

for the CIR model.[6] This further explains why the values of the dividend and the shareholder option based on the CIR model are anyhow comparable with the ones based on the HW or the G2++ model.

Indeed, also the mean standard deviation of the 10 year ZCB prices in the HW model deviates from the G2++ model, but this happens much slower than in case of the CIR model. Turning back, we find the results in Table 5.14 confirming this trend, the value of the interest rate guarantee is lower for the HW model than for the G2++ model.

The values of the interest rate guarantee can also be regarded as some measure of risk from the perspective of the insurer. The higher the guarantee for any portfolio is, the higher the risk the insurer bears. On the other hand, taking a higher risk commonly demands a higher expected return. And, indeed, for any particular set of surplus parameters α and β, the values of the dividend option are, by implication of higher asset returns, the higher the values of the interest rate guarantees are. The respective results can be seen in Table 5.15.

One could have expected analogous results for the shareholder option, but analysing the results in Table 5.16 disproves this. For any particular set of surplus parameters α and β, the HW model, for example, has a lower value of the interest rate guarantee and a lower value of the dividend option, but the value of the respective shareholder option is anyhow greater than for the G2++ model. This can be traced back to the mechanism determining the policyholder profit that contains a

[6]As shown on Tables 5.1–5.4, the technical interest rates are between 2.0% and 3.5%, whereas the mean annualised return on a ZCB of any maturity in the CIR model climbs above the maximum rate of 3.5% after about 8 years of simulation. Confer Fig. 3.15 therefor.

buffering element. Therefore these results provide important insights to understand the functionality of the bonus distribution machanism of the entire model.

The current analysis broaches the issue of interest rate model risk. After all, we can conclude that the CIR model is unsuitable for the purpose of this modelling approach as it significantly underestimates the interest rate risk in terms of the low values of the interest rate guarantee. Furthermore, the quantitative differences between the HW model and the G2++ model are identified. In terms of the higher total values of all the embedded options, the G2++ is a more conservative model than the HW model. Together with the theoretical background and the description of the calibration process given in Chap. 3, the user should be able to decide whether to use the HW or the G2++ approach for his own purposes.

Next we shall examine different asset allocations. Within the basic simulation run, a realistic composition of the reference portfolio consisting of an equity fund, a property fund and ZCBs, is assumed. In the subsequent section there will be an analysis of a more progressive and a more conservative asset portfolio.

Interest rate guarantee/variation of the stochastic scenario file for ZCB prices

Table 5.14 Values of the interest rate guarantee in Swiss Francs using different stochastic scenario files for ZCB prices. The values in brackets denote the relative difference to the basic simulation run

Component	Surplus parameter combination		
Single premium	$\alpha = 0.8, \beta = 0.10$	$\alpha = 0.4, \beta = 0.10$	$\alpha = 0.4, \beta = 0.05$
CIR model	4,378,597	4,288,742	4,520,970
	(−0.8891)	(−0.8830)	(−0.8849)
HW model	31,465,517	29,006,597	31,364,902
	(−0.2033)	(−0.2086)	(−0.2017)
G2++ model	**39,497,246**	**36,653,299**	**39,288,592**
	(0.0000)	(0.0000)	(0.0000)
Regular premium	$\alpha = 0.8, \beta = 0.10$	$\alpha = 0.4, \beta = 0.10$	$\alpha = 0.4, \beta = 0.05$
CIR model	2,652,690	1,831,719	3,293,554
	(−0.9379)	(−0.9455)	(−0.9245)
HW model	39,470,813	30,777,100	40,927,361
	(−0.0753)	(−0.0840)	(−0.0620)
G2++ model	**42,686,809**	**33,598,175**	**43,633,497**
	(0.0000)	(0.0000)	(0.0000)

Dividend option/variation of the stochastic scenario file for ZCB prices

Table 5.15 Values of the dividend option in Swiss Francs using different stochastic scenario files for ZCB prices. The values in brackets denote the relative difference to the basic simulation run

Component	Surplus parameter combination		
Single premium	$\alpha = 0.8, \beta = 0.10$	$\alpha = 0.4, \beta = 0.10$	$\alpha = 0.4, \beta = 0.05$
CIR model	80,432,894	81,316,528	80,350,689
	(−0.1013)	(−0.0728)	(−0.1013)
HW model	87,323,784	85,931,504	87,263,243
	(−0.0243)	(−0.0201)	(−0.0240)
G2++ model	**89,501,240**	**87,696,664**	**89,404,943**
	(0.0000)	(0.0000)	(0.0000)
Regular premium	$\alpha = 0.8, \beta = 0.10$	$\alpha = 0.4, \beta = 0.10$	$\alpha = 0.4, \beta = 0.05$
CIR model	734,337,932	728,682,495	743,899,301
	(−0.1784)	(−0.1709)	(−0.1747)
HW model	879,668,991	865,393,444	886,808,581
	(−0.0158)	(−0.0153)	(−0.0162)
G2++ model	**893,824,435**	**878,842,480**	**901,393,649**
	(0.0000)	(0.0000)	(0.0000)

Shareholder option/variation of the stochastic scenario file for ZCB prices

Table 5.16 Values of the shareholder option in Swiss Francs using different stochastic scenario files for ZCB prices. The values in brackets denote the relative difference to the basic simulation run

Component	Surplus parameter combination		
Single premium	$\alpha = 0.8, \beta = 0.10$	$\alpha = 0.4, \beta = 0.10$	$\alpha = 0.4, \beta = 0.05$
CIR model	25,096,438	25,499,424	24,558,054
	(0.367)	(0.3005)	(0.3606)
HW model	18,695,497	19,919,154	18,375,137
	(0.0184)	(0.0159)	(0.0181)
G2++ model	**18,358,262**	**19,607,117**	**18,048,894**
	(0.0000)	(0.0000)	(0.0000)
Regular premium	$\alpha = 0.8, \beta = 0.10$	$\alpha = 0.4, \beta = 0.10$	$\alpha = 0.4, \beta = 0.05$
CIR model	123,587,076	128,345,575	118,949,311
	(0.2061)	(0.1489)	(0.1678)
HW model	104,924,167	113,790,643	104,003,838
	(0.0240)	(0.0186)	(0.0210)
G2++ model	**102,467,939**	**111,707,333**	**101,860,543**
	(0.0000)	(0.0000)	(0.0000)

5.4 Variation of the Asset Allocation

In general we have three different investment instruments where the initial amount of the asset portfolio A_0 can be invested in: An equity fund, a property fund and ZCBs. Based on available asset data particular weights w_i may be determined in order to compose a realistic mapping of Helvetia's current asset portfolio. Recall the weights outlined at the beginning of this chapter.

Instead of the realistic asset allocation, that is de facto the observed asset allocation, we shall also consider a more "progressive" and a more "conservative" asset allocation. In plain language, a more progressive asset allocation means that, compared to the realistic asset allocation, equity and property are weighted stronger. In the opposite case of a more conservative asset allocation, it is the long term ZCBs that are weighted stronger.

To define the progressive asset allocation, the portfolio weights for the equity fund and the property fund are multiplied by a factor 2 and 1.5 respectively. The weights for the ZCBs get scaled proportionally such that it is a feasible asset allocation again, meaning that the sum of all weights is equal to 1. Similarly, the conservative asset allocation is defined. Based on the realistic asset allocation, the weights for the 10 year ZCB, the 15 year ZCB and the 20 year ZCB are multiplied by a factor 1.25, whereas the other weights get scaled proportionally such that the sum of all weights is equal to 1. The resulting weights can be found in Table 5.17.

Table 5.17 Portfolio weights to define the progressive, the realistic and the conservative asset allocation

Instrument	Weight		
	Progressive	Relistic	Conservative
Equity fund	0.0956	0.0478	0.0249
Property fund	0.2552	0.1701	0.0885
1 year ZCB	0.0349	0.0421	0.0219
3 year ZCB	0.0124	0.0149	0.0078
5 year ZCB	0.0563	0.0678	0.0353
10 year ZCB	0.1685	0.2030	0.2538
15 year ZCB	0.1280	0.1542	0.1928
20 year ZCB	0.2491	0.3001	0.3750
Total	1.0000	1.0000	1.0000

In practice the composition of an insurance company's asset portfolio is a very delicate issue and is never done that simple as I do within this sensitivity analysis. Naturally there are also legal aspects to be taken into account when building an asset portfolio. I will not further consider this issue here and therefore the interested reader is advised to study FINMA (2008) that provides investment guidelines based on legal foundations for Swiss insurers. Above, the determination of the new weights happens arbitrarily and should therefore be borne in mind when considering the results shown in Tables 5.18, 5.19 and 5.20.

The results from the various asset allocations are quite impressive from two different angles. First of all, we constitute the big spread of the values across all types of options. Something that hardly could have been predicted.

Second, we realise that the progressive asset allocation performs lower values for the interest rate guarantee and higher values for the dividend and the shareholder option than the realistic allocation does. Hence, this is a clear statement to the insurer to implement a more progressive asset allocation strategy. But beware, this suggestion is only feasible as long as we trust in the underlying model assumptions and its calibration. Furthermore, in practice there might be some limitations, too. As mentioned above the (insurance) supervisory authority might release regulations on the decomposition of the asset portfolio or the insurer has to sell part of its short term assets due to temporary financial distress.

In Sect. 3.7 I have already defined that we are, for the purpose of this modelling approach, operating in a perfect market where transaction costs, liquidity squeezes, tax effects and other market imperfections may be ignored. But there is one important issue that has not been discussed so far. In practice, any larger asset portfolio is hedged to a certain extent by various hedging instruments like, for example, stock options. In order to form an even more realistic modelling approach the design of a hedging strategy could be an additional issue.

Application of various asset allocation strategies finally demonstrates how sensitive this model responds to this input. For sure, the results are impressive but anyhow have to be interpreted carefully. Nevertheless, this sensitivity analysis enables a reasonable and powerful basis to perform ALM actions. The following section presents a historic comparison by performing a basic simulation run using as at the end of 2004 data.

Interest rate guarantee/variation of the stochastic scenario file for ZCB prices

Table 5.18 Values of the interest rate guarantee in Swiss Francs using different asset allocations. The values in brackets denote the relative difference to the basic simulation run

Component	Surplus parameter combination		
Single premium	$\alpha = 0.8, \beta = 0.10$	$\alpha = 0.4, \beta = 0.10$	$\alpha = 0.4, \beta = 0.05$
Progressive	9,575,219	8,800,706	9,384,357
	(−0.7576)	(−0.7599)	(−0.7611)
Realistic	**39,497,246**	**36,653,299**	**39,288,592**
	(0.0000)	(0.0000)	(0.0000)
Conservative	110,060,831	101,956,309	109,717,585
	(1.7865)	(1.7816)	(1.7926)
Regular premium	$\alpha = 0.8, \beta = 0.10$	$\alpha = 0.4, \beta = 0.10$	$\alpha = 0.4, \beta = 0.05$
Progressive	15,620,602	11,303,350	15,938,335
	(−0.6341)	(−0.6636)	(−0.6347)
Realistic	**42,686,809**	**33,598,175**	**43,633,497**
	(0.0000)	(0.0000)	(0.0000)
Conservative	93,891,681	74,006,432	92,581,790
	(1.1995)	(1.2027)	(1.1218)

Dividend option/variation of the stochastic scenario file for ZCB prices

Table 5.19 Values of the dividend option in Swiss Francs using different asset allocations. The values in brackets denote the relative difference to the basic simulation run

Component	Surplus parameter combination		
Single premium	$\alpha = 0.8, \beta = 0.10$	$\alpha = 0.4, \beta = 0.10$	$\alpha = 0.4, \beta = 0.05$
Progressive	163,654,590	166,927,493	164,256,761
	(0.8285)	(0.9035)	(0.8372)
Realistic	**89,501,240**	**87,696,664**	**89,404,943**
	(0.0000)	(0.0000)	(0.0000)
Conservative	62,368,509	55,457,057	61,928,271
	(−0.3032)	(−0.3676)	(−0.3073)
Regular premium	$\alpha = 0.8, \beta = 0.10$	$\alpha = 0.4, \beta = 0.10$	$\alpha = 0.4, \beta = 0.05$
Progressive	1,224,557,435	1,216,310,160	1,231,962,871
	(0.3700)	(0.3840)	(0.3667)
Realistic	**893,824,435**	**878,842,480**	**901,393,649**
	(0.0000)	(0.0000)	(0.0000)
Conservative	691,646,721	667,230,532	694,137,805
	(−0.2262)	(−0.2408)	(−0.2299)

Shareholder option/variation of the stochastic scenario file for ZCB prices

Table 5.20 Values of the shareholder option in Swiss Francs using different asset allocations. The values in brackets denote the relative difference to the basic simulation run

Component	Surplus parameter combination		
Single premium	$\alpha = 0.8, \beta = 0.10$	$\alpha = 0.4, \beta = 0.10$	$\alpha = 0.4, \beta = 0.05$
Progressive	23,913,451	24,654,677	23,825,698
	(0.3026)	(0.2574)	(0.3201)
Realistic	**18,358,262**	**19,607,117**	**18,048,894**
	(0.0000)	(0.0000)	(0.0000)
Conservative	13,097,616	14,710,474	12,895,575
	(−0.2866)	(−0.2497)	(−0.2855)
Regular premium	$\alpha = 0.8, \beta = 0.10$	$\alpha = 0.4, \beta = 0.10$	$\alpha = 0.4, \beta = 0.05$
Progressive	123,684,391	130,468,140	123,349,009
	(0.2071)	(0.1679)	(0.2110)
Realistic	**102,467,939**	**111,707,333**	**101,860,543**
	(0.0000)	(0.0000)	(0.0000)
Conservative	74,864,288	85,673,590	75,900,270
	(−0.2694)	(−0.2331)	(−0.2549)

5.5 Analysis of as at the End of 2004 Data

The numerical analysis that is performed in this thesis is based on a series of stochastic simulation runs. Its idea is to construct a mathematical apparatus to forecast the economic future from which we can derive the today's values of selected embedded options. The crucial element of this stochastic simulation is the underlying data basis that quasi dictates the future within the model. And in turn, this data basis consists of current data like the swap rate curve for discounting or the policy data providing the insurance specific information. Partially, we also make use of past data when, for example, estimating the mean expected return of the share price and the property price index.[7]

Naturally, all the time dependent parameters used for calculation or calibration change as time goes by and so the results do. To have a comparison over time, there is a basic simulation run in the current section using consequently as at the end of 2004 data. The only data that could not be found exactly as at this date were the swap rates. Since Datastream provides swap rates for Swiss Francs only as of 28 February 2005 for particular maturities, this rates are taken for all the considered maturities as they were as at the end of 2004. To remain consistent, also the swaption implied volatilities for calibrating the G2++ model are collected as at this date.

Concretely, for the end of 2004 basic simulation run the following quantities have to be edited:

- Policy data including assumptions on the expenses collected as at the end of 2004,
- Mortality table EKM/F_96_00_II to determine the cash flow scheme,
- Swap rates as at 28 February 2005 for discounting and to calibrate the G2++ model,
- Swaption implied volatilities as at 28 February 2005 to calibrate the G2++ model and
- SPI and IAZI index data between 1 January 2000 and 31 December 2004 to calibrate the share and the property price process.

The interested reader may find the data used for calibrating the various processes as at the end of 2004 in Appendix A.2. All the rest of the assumptions is borrowed from the end of 2009 basic simulation run as outlined at the beginning of the current chapter.

Even though we are actually going to consider notionally the same portfolios of insurance, there are 5 years in between and so the portfolios have changed. Policies might have matured, were terminated due to the death of the policyholder or, for the latest tariff generation, new policies might have been issued. However, before performing the end of 2004 basic simulation run, it makes sense to have a look at

[7] Confer Sect. 3.3.1.

the initial values for the asset portfolio A, the policy account P and the buffer B. Find the results in Table 5.21 for the end of 2004 (eo 2004) and the end of 2009 (eo 2009) data.

Table 5.21 Historic comparison for the initial balance sheet entries A_0, P_0 and B_0 providing the source for the basic simulation runs as at the end of 2004 (eo 2004) and 2009 (eo 2009). Amounts of money are denoted in Swiss Francs

Category	No. pols	A_0	P_0	B_0
Endowment insurance, single premium				
Eo 2004	32,346	1,457,654,788	1,460,640,960	−2,986,171
Eo 2009	24,112	1,082,494,476	1,105,993,704	−23,499,229
Endowment insurance, regular premium				
Eo 2004	66,430	1,249,252,358	998,692,904	250,559,453
Eo 2009	72,246	1,558,950,089	1,367,346,236	191,603,853
Whole life insurance, single premium				
Eo 2004	1,212	63,407,857	52,856,577	10,551,279
Eo 2009	1,045	53,641,355	47,701,536	5,939,819
Whole life insurance, regular premium				
Eo 2004	1,120	11,858,110	10,189,710	1,668,400
Eo 2009	930	11,780,734	10,501,120	1,279,613
Term life insurance, regular premium				
Eo 2004	12,746	35,017,241	−21,994,884	57,012,125
Eo 2009	11,181	34,442,838	−14,640,656	49,083,496
Pure endowment insurance, single premium				
Eo 2004	1329	63,868,202	66,229,889	−2,361686
Eo 2009	550	21,528,433	22,855,272	−1,326,841
Pure endowment insurance, regular premium				
Eo 2004	1,605	36,448,682	38,404,196	−1,955,514
Eo 2009	1,373	37,706,027	41,254,858	−3,548,829

The most remarkable observation we can make is that, except for single premium pure endowment insurance, the absolute values of the buffer have worsened since end of 2004. There exist several factors that give reasons for this. Neglecting the fact that we are considering absolute values even though we have different numbers of policies, the point is that these initial values are derived for totally different points in time. Five years ago, the swap rates were higher, the assumptions on expenses were different and, most notably, the insurance portfolios were about "five years younger".[8] After all, this means that for the eo 2004 data the market consistent

[8]Confer Fig. A.4 in Sect. A.2 of the Appendix for a comparison of the swap rates.

valuation, that is the discounting with the market interest rates in particular, acts even longer on the cash flows, relative to the total portfolio duration. Through this, it becomes evident that the market consistent valuation methodology reaches its limits here. In fact, the problem is not the methodology itself, but the mixing with another valuation methodology, the statutory valuation methodology.[9] And this leads to an inconsistency within the model for comparisons of different calculation dates.

A priori it is not that clear what results shall be expected regarding the values of the embedded options. On the one hand we see that the initial values remarkably differ but, on the other hand, the application of the modelling approach is anyhow consistent in a sense that the methodology is applied identically for both points in time and so one could expect potentially similar results. However, due to the fact that the number of policies is different for the two sets of data, it would not make sense to consider any absolute values. Therefore, for the purpose of this particular analysis, we will consider the option's values relative to the value of the initial asset portfolio A_0. Find the results in Tables 5.22–5.24.

In consequence of the mostly higher initial buffer values B_0, the values of the shareholder option for the eo 2004 run are, as shown in Table 5.24, constantly higher than for the eo 2009 run. Much more interesting are the results for the dividend option as presented in Table 5.23. After having looked at the results of the shareholder option, one could have expected similar results in a sense that they have increased for the eo 2004 run, too. But in fact, they have massively decreased. Indeed, the parameters resulting from the eo 2004 calibration for the share, the property and the ZCB price process indicate that the total return on the asset portfolio A might be lower than in the eo 2009 case and so a lower value for the dividend option should actually not surprise in general.[10] However, in line with the results of the shareholder option, the order of which the results for the dividend option differ from the eo 2009 run, definitely raises some doubts about this. Obviously there are even greater discrepancies as mentioned above in connection with the initial buffer values, when applying both, a market consistent and a statutory valuation methodology.

There is another argument that maintains this finding. When considering the results for the interest rate guarantee in Table 5.22 we constitute a change of sign, the values of the interest rate guarantee for single premium portfolios are relatively lower in case of the eo 2004 basic simulation run, but are higher for regular premium portfolios. Based on the current knowledge on this modelling approach this is definitely at odds with the values of the other options, where the deviations of the eo 2004 run to the eo 2009 run do not show a change of sign between premium payment types.

[9]By statutory valuation I mean the traditional approach the disclosed policy reserves and the surplus reserves, which are used as the initial value of the asset portfolio A_0, are calculated.

[10]Confer the calibration data in Sect. A.2 of the Appendix.

The market consistent valuation of embedded options is clearly a challenging issue. Naturally their values change over time as the underlying market data change. But the current analysis has impressively demonstrated where the limitations of the market consistent valuation methodology are when applying a modelling approach mixing both, market consistent and statutory valuation methodology.

This is the point I deduce one fundamental and possibly provocative suggestion: Throughout this thesis the today's importance of market consistent valuation was emphasised. As we compare nowadays solvency standards or valuation frameworks, it gets obvious that market consistent valuation principles are on the way to replace traditional valuation principles.[11] Aside from this, so far the empirical analysis shows, amongst other things, an apparent conflict due to the application of different valuation principles like statutory and market consistent. This means that insurers should successively move towards a comprehensive and entirely consistent application of the market consistent methodologies regarding valuation, reserving and, as far as possible, also pricing. In my opinion this is the corollary to get even more meaningful values for the embedded options in particular and to attain an entirely consistent risk management in general. Indeed, this is quite a strong argument but, from the perspective of research, a legitimate request. Notice that this suggestion is not only valid for life insurers but for the whole insurance industry.

The sensitivity analysis of the subsequent section replaces the swap rate curve by the SST interest rate curve, that are the *zero coupon rates*.

Interest rate guarantee/analysis of as at the end of 2004 data

Table 5.22 Values of the interest rate guarantee relative to the value of the initial asset portfolio A_0 using as at the end of 2004 (eo 2004) data. The values in brackets denote the relative difference to the basic simulation run as at the end of 2009 (eo 2009)

Component	Surplus parameter combination		
Single premium	$\alpha = 0.8, \beta = 0.10$	$\alpha = 0.4, \beta = 0.10$	$\alpha = 0.4, \beta = 0.05$
Eo 2004	0.0217	0.0204	0.0232
	(−0.3632)	(−0.3545)	(−0.3167)
Eo 2009	**0.0341**	**0.0317**	**0.0339**
	(0.0000)	(0.0000)	(0.0000)
Regular premium	$\alpha = 0.8, \beta = 0.10$	$\alpha = 0.4, \beta = 0.10$	$\alpha = 0.4, \beta = 0.05$
Eo 2004	0.0360	0.0310	0.0391
	(0.3861)	(0.5144)	(0.4725)
Eo 2009	**0.0260**	**0.0205**	**0.0266**
	(0.0000)	(0.0000)	(0.0000)

[11] Confer the description of SST, Solvency II and MCEV presented in Chap. 1.

Dividend option/analysis of as at the end of 2004 data

Table 5.23 Values of the dividend option relative to the value of the initial asset portfolio A_0 using as at the end of 2004 (eo 2004) data. The values in brackets denote the relative difference to the basic simulation run as at the end of 2009 (eo 2009)

Component	Surplus parameter combination		
Single premium	$\alpha = 0.8, \beta = 0.10$	$\alpha = 0.4, \beta = 0.10$	$\alpha = 0.4, \beta = 0.05$
Eo 2004	0.0376	0.0354	0.0400
	(−0.5131)	(−0.5327)	(−0.4821)
Eo 2009	**0.0773**	**0.0758**	**0.0772**
	(0.0000)	(0.0000)	(0.0000)
Regular premium	$\alpha = 0.8, \beta = 0.10$	$\alpha = 0.4, \beta = 0.10$	$\alpha = 0.4, \beta = 0.05$
Eo 2004	0.3663	0.3520	0.3753
	(−0.3266)	(−0.3420)	(−0.3160)
Eo 2009	**0.5441**	**0.5349**	**0.5487**
	(0.0000)	(0.0000)	(0.0000)

Shareholder option/analysis of as at the end of 2004 data

Table 5.24 Values of the shareholder option relative to the value of the initial asset portfolio A_0 using as at the end of 2004 (eo 2004) data. The values in brackets denote the relative difference to the basic simulation run as at the end of 2009 (eo 2009)

Component	Surplus parameter combination		
Single premium	$\alpha = 0.8, \beta = 0.10$	$\alpha = 0.4, \beta = 0.10$	$\alpha = 0.4, \beta = 0.05$
Eo 2004	0.0179	0.0187	0.0167
	(0.1258)	(0.1043)	(0.0739)
Eo 2009	**0.0159**	**0.0169**	**0.0156**
	(0.0000)	(0.0000)	(0.0000)
Regular premium	$\alpha = 0.8, \beta = 0.10$	$\alpha = 0.4, \beta = 0.10$	$\alpha = 0.4, \beta = 0.05$
Eo 2004	0.0727	0.0800	0.0695
	(0.1663)	(0.1760)	(0.1202)
Eo 2009	**0.0624**	**0.0680**	**0.0620**
	(0.0000)	(0.0000)	(0.0000)

5.6 Variation of the Market Interest Rates

As suggested for the MCEV and the Solvency II framework, exclusively the current swap rate curve is applied for discounting in conjunction with this modelling approach.[12] Instead, within the SST framework another interest rate curve is applied,

[12]Confer principle 14 in CFO Forum (2009a) and annex F in CEIOPS (2010b).

the *zero coupon rates*. These rates as at the end of 2009 for Swiss Francs are derived from Swiss Government bonds and deemed to be risk-free. They can be found in the so called *SST Template 2009* that is available on the FINMA webpage.[13] Currently there is a lively discussion around the SST whether to go on using the zero coupon rates for discounting or to adapt to the international standards like, for example, Solvency II using the swap rates. Keller et al. (2011) recently published a brief statement that confirms the use of the zero coupon rates within the SST from an actuarial point of view.

I will use the opportunity to perform a simulation run applying the zero coupon rates from the SST instead of the so far used swap rates.

Below Fig. 5.2 plots both interest rate curves, the swap rate curve and the zero coupon curve as at the end of 2009.

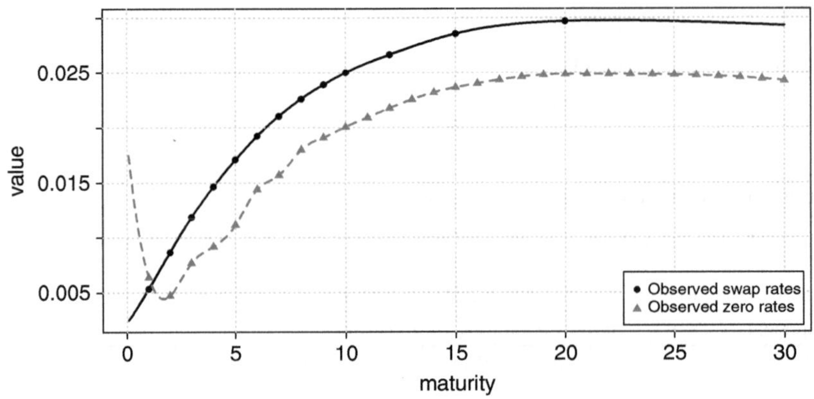

Fig. 5.2 The current swap rate curve and the current zero coupon curve for Swiss Francs as at 31.12.2009. The values between the observed rates were interpolated performing the cubic spline procedure `spline(...)` in **R** (Source: Datastream (swap rates)/FINMA (zero coupon rates))

From Fig. 5.2 we observe that the zero coupon rates are on average approximately 50 basis points below the swap rates. What consequences can we expect from this? First of all, remember that the initial value of the policy account P_0 is defined as the present value of a portfolio's net cash flow. In most of the cases, P_0 is positive when discounting with the swap rate curve and so we can assume that P_0 increases if we discount with the lower zero coupon rates. This further leads to a lower initial buffer B_0.

We also have to keep in mind that not only the initial values of the policy account P_0 and the buffer B_0 are affected by the change of the interest rate curve. As we see on page 95, the policy account before bonus calculations at the end of period t, $P^{bb}(t^+)$, gets multiplied by the factor $(1/P(t,t^+))$, whereas $P(t,t^+)$ is derived from the currently used interest rate curve.

[13]See http://www.finma.ch/e/pages/default.aspx. Last checked: 1 June 2011.

Lastly, recall the Formulae 4.11, 4.12 and 4.13 to see how the embedded options are defined. In actuarial terms these definitions are nothing but present values of particular cash flows. Hence lower discount rates might yield higher present values given the terminal values are the same.

We constitute that replacing the underlying interest rate curve has a number of impacts and so it is quite difficult to predict what changes will occur. Therefore, see the results in Tables 5.25, 5.26 and 5.27.

Surprisingly the values of the interest rate guarantee do not show a homogeneous trend in Table 5.25 when considering the results for single and regular premium policies. However, we see lower total values for the interest rate guarantee when assuming the zero coupon rates.

In case of the values of the dividend and the shareholder option it is the opposite way around as shown in Tables 5.26 and 5.27. The lower interest rates yield higher total values for the embedded options.

For the purpose of this section it is sufficient to recognise what impacts the application of different discounting factors has on the values of the embedded options *by trend*. We can conclude that assuming a lower interest rate curve yields a higher present value of the policy account and thus reflects a more conservative behaviour in the interest rate assumptions but anyhow produces lower total values for the interest rate guarantee and higher total values for the dividend and the shareholder option.

The next section performs a "what-if analysis" by assuming premium reduction instead of an interest-bearing accumulation as the surplus distribution mechanism.

Interest rate guarantee/variation of the market interest rates

Table 5.25 Values of the interest rate guarantee in Swiss Francs using using different market interest rates. The values in brackets denote the relative difference to the basic simulation run

Component	Surplus parameter combination		
Single premium	$\alpha = 0.8, \beta = 0.10$	$\alpha = 0.4, \beta = 0.10$	$\alpha = 0.4, \beta = 0.05$
Zero coupon rates	42,057,950	40,182,190	41,685,500
	(0.0648)	(0.0963)	(0.0610)
Swap rates	**39,497,246**	**36,653,299**	**39,288,592**
	(0.0000)	(0.0000)	(0.0000)
Regular premium	$\alpha = 0.8, \beta = 0.10$	$\alpha = 0.4, \beta = 0.10$	$\alpha = 0.4, \beta = 0.05$
Zero coupon rates	34,354,591	26,662,384	35,391,234
	(−0.1952)	(−0.2064)	(−0.1889)
Swap rates	**42,686,809**	**33,598,175**	**43,633,497**
	(0.0000)	(0.0000)	(0.0000)

Dividend option/variation of the market interest rates

Table 5.26 Values of the dividend option in Swiss Francs using different market interest rates. The values in brackets denote the relative difference to the basic simulation run

Component	Surplus parameter combination		
Single premium	$\alpha = 0.8, \beta = 0.10$	$\alpha = 0.4, \beta = 0.10$	$\alpha = 0.4, \beta = 0.05$
Zero coupon rates	97,986,909	98,169,191	98,051,216
	(0.0948)	(0.1194)	(0.0967)
Swap rates	**89,501,240**	**87,696,664**	**89,404,943**
	(0.0000)	(0.0000)	(0.0000)
Regular premium	$\alpha = 0.8, \beta = 0.10$	$\alpha = 0.4, \beta = 0.10$	$\alpha = 0.4, \beta = 0.05$
Zero coupon rates	980,609,268	976,752,076	992,352,473
	(0.0971)	(0.1114)	(0.1009)
Swap rates	**893,824,435**	**878,842,480**	**901,393,649**
	(0.0000)	(0.0000)	(0.0000)

Shareholder option/variation of the market interest rates

Table 5.27 Values of the shareholder option in Swiss Francs using different market interest rates. The values in brackets denote the relative difference to the basic simulation run

Component	Surplus parameter combination		
Single premium	$\alpha = 0.8, \beta = 0.10$	$\alpha = 0.4, \beta = 0.10$	$\alpha = 0.4, \beta = 0.05$
Zero coupon rates	18,243,825	19,206,140	18,059,838
	(−0.0062)	(−0.0205)	(0.0006)
Swap rates	**18,358,262**	**19,607,117**	**18,048,894**
	(0.0000)	(0.0000)	(0.0000)
Regular premium	$\alpha = 0.8, \beta = 0.10$	$\alpha = 0.4, \beta = 0.10$	$\alpha = 0.4, \beta = 0.05$
Zero coupon rates	113,480,375	121,086,908	112,272,833
	(0.1075)	(0.0840)	(0.1022)
Swap rates	**102,467,939**	**111,707,333**	**101,860,543**
	(0.0000)	(0.0000)	(0.0000)

5.7 Variation of the Surplus Distribution Mechanism

Remember the portfolio cash flow scheme that forms the heart of this modelling algorithm. Due to this construction, certain information on individual policies get lost. So, for example, it is not possible to take the exact surplus distribution mechanism into account. As explained in Sect. 4.2, the modelling approach is simplified by assuming that for all policies the same mechanism is applied. Since it is the vast majority of the policies that has the interest-bearing accumulation as the surplus distribution mechanism, this type is chosen as the default case.

Within this sensitivity analysis I will take the opportunity to perform a "what-if analysis" by assuming premium reduction instead of the interest-bearing accumulation. Naturally, this makes sense only for policies that are paid for regular premiums. Find the results in Tables 5.28, 5.29 and 5.30.

Paying surpluses as premium reduction actually means that a possible surplus amount is allocated according to the premium income scheme and no longer according to the benefit outgo scheme as in the case of an interest-bearing accumulation. And since premiums usually come in earlier than death outgo and maturity outgo become due, the policyholder also gets its surplus earlier. Thus, one could expect, due to the discounting, that the dividend option slightly increases when premium reductions are assumed. But, observing Table 5.29, the dividend option has remarkably decreased and so did the interest rate guarantee as shown in Table 5.28. What is the reason for this? Since the amount of surplus to distribute is determined independent of the distribution mechanism, it must be the distribution mechanism itself that makes the difference.

The pattern of premium income is clearly different to the pattern of death and maturity outgo. By trend premium income decreases, unlike death and maturity outgo that decrease. Consequently, the ratio that controls the premium reduction to be paid out in the following period is higher than the respective ratio for the interest-bearing accumulation.[14] Therefore, a higher payout ratio at start reduces the total amount of the undistributed surplus reserve and this again reduces the total returns on it. After all, this is the explanation for the decrease of the values of the dividend options in case of premium reductions comes with a lower value of the dividend option.

The values of the shareholder option are shown in Table 5.30. We constitute that the impacts on these values, whether premium reduction or interest-bearing accumulation is applied, are fairly small. But, as outlined above already, the change of the surplus distribution mechanism does not directly change the amount of surplus that is determined to be distributed and so the shareholder profits are only marginally affected.

After having analysed different surplus *distribution* mechanisms in the current section, Sect. 5.8 presents various ways to *determine* surpluses.

[14]Confer the modelling algorithm on page 95 to see the definition of the ratios.

Interest rate guarantee/variation of the surplus distribution mechanism

Table 5.28 Values of the interest rate guarantee in Swiss Francs using different surplus distribution mechanisms. The values in brackets denote the relative difference to the basic simulation run

Component	Surplus parameter combination		
Regular premium	$\alpha = 0.8, \beta = 0.10$	$\alpha = 0.4, \beta = 0.10$	$\alpha = 0.4, \beta = 0.05$
Premium reduction	38,365,830	30,375,511	39,983,281
	(−0.1012)	(−0.0959)	(−0.0837)
Accumulation	**42,686,809**	**33,598,175**	**43,633,497**
	(0.0000)	(0.0000)	(0.0000)

Dividend option/variation of the surplus distribution mechanism

Table 5.29 Values of the dividend option in Swiss Francs using different surplus distribution mechanism. The values in brackets denote the relative difference to the basic simulation run

Component	Surplus parameter combination		
Regular premium	$\alpha = 0.8, \beta = 0.10$	$\alpha = 0.4, \beta = 0.10$	$\alpha = 0.4, \beta = 0.05$
Premium reduction	777,266,761	774,954,397	787,104,595
	(−0.1304)	(−0.1182)	(−0.1268)
Accumulation	**893,824,435**	**878,842,480**	**901,393,649**
	(0.0000)	(0.0000)	(0.0000)

Shareholder option/variation of the surplus distribution mechanism

Table 5.30 Values of the shareholder option in Swiss Francs using different surplus distribution mechanisms. The values in brackets denote the relative difference to the basic simulation run

Component	Surplus parameter combination		
Regular premium	$\alpha = 0.8, \beta = 0.10$	$\alpha = 0.4, \beta = 0.10$	$\alpha = 0.4, \beta = 0.05$
Premium reduction	102,895,683	111,931,556	101,730,519
	(0.0042)	(0.002)	(−0.0013)
Accumulation	**102,467,939**	**111,707,333**	**101,860,543**
	(0.0000)	(0.0000)	(0.0000)

5.8 Variation of the Surplus Determination

This section is about analysing various ways to determine surpluses. Haberman et al. (2003) discuss different definitions of their policy interest rate r_P, but due to different modelling approaches, they cannot be adopted here.[15] Instead I suggest the following.

[15]See the summary of the cited paper presented in the literature survey in Sect. 2.2.

Remember the modelling algorithm on page 95 where the surplus amounts for shareholders and policyholders are defined sequentially.

V0/the default case

$$\text{SH\_PROFIT}(t^+) = \max\{0, \min\{B^{bb}(t^+), \text{sh\_margin} \cdot P^{bb}(t^+)\}\}$$
$$\text{PH\_PROFIT}(t^+) = \max\{0, \alpha \cdot (B^{bb}(t^+) - \text{SH\_PROFIT}(t^+) - \beta P^{bb}(t^+))\}$$

This is the default case denoted by V0. Here the order implies that shareholders are served prior to policyholders. Instead, one could also assume the opposite case and simply interchange this order. This yields in the following definitions that shall be denoted by V1.

V1/interchange of the order

$$\text{PH\_PROFIT}(t^+) = \max\{0, \alpha \cdot (B^{bb}(t^+) - \beta P^{bb}(t^+))\}$$
$$\text{SH\_PROFIT}(t^+) = \max\{0, \min\{B^{bb}(t^+) - \text{PH\_PROFIT}(t^+), \text{sh\_margin} \cdot P^{bb}(t^+)\}$$

There shall be a second modification, denoted by V2, that redefines the policyholder profit by the following term.

V2/alternative definition of the policyholder profit

$$\text{SH\_PROFIT}(t^+) = \max\{0, \min\{B^{bb}(t^+), \text{sh\_margin} \cdot P^{bb}(t^+)\}\}$$
$$\text{PH\_PROFIT}(t^+) = \max\{0, \min(B^{bb}(t^+) - \text{SH\_PROFIT}(t^+), d \cdot \beta \cdot P^{bb}(t^+))\}$$

The structure the policyholder profit is defined here is derived from the shareholder profit and thus the crucial point is that the former buffer property gets alleviated. However, note that V2 is rather a heuristic attempt in order to vary the surplus determination and comes without any practical foundation. For this reason, the parameter d in the definition above was set arbitrarily to 0.2. However, this shall be seen as a motivation to further adjust the surplus determination by customising for company specific reasons.

The results are shown in the Tables 5.31, 5.32 and 5.33.

We see that interchanging the order of shareholder profit and policyholder profit (V1) has only little impact on the values of the options. The values of the dividend option slightly increase whereas the values of the shareholder option decrease compared to the default case (V0). Also, and this is probably the most notable point for this sensitivity, the values of the interest rate guarantees moderately increase. Still this analysis is rather poorly meaningful.

The alternative definition of the policyholder profit (V2) was deemed to be a heuristic attempt to further improve this definition, and I have explained that there is no immediate practical foundation for this. This means that the parameter d (and also β) was chosen completely arbitrarily. Moreover, it is difficult to compare the resulting values with the default case since the definitions of the policyholder profits structurally differ. For this reason, there shall be no comment added to these results and its interpretation is left to the reader.

Interest rate guarantee/variation of the surplus determination

Table 5.31 Values of the interest rate guarantee in Swiss Francs using different surplus determination. The values in brackets denote the relative difference to the basic simulation run

Component	Surplus parameter combination		
Single premium	$\alpha = 0.8, \beta = 0.10$	$\alpha = 0.4, \beta = 0.10$	$\alpha = 0.4, \beta = 0.05$
V0	**39,497,246**	**36,653,299**	**39,288,592**
	(0.0000)	(0.0000)	(0.0000)
V1	39,928,461	36,804,430	39,763,041
	(0.0109)	(0.0041)	(0.0121)
V2	40,708,463	40,708,463	37,408,293
	(0.0307)	(0.1106)	(−0.0479)
Regular premium	$\alpha = 0.8, \beta = 0.10$	$\alpha = 0.4, \beta = 0.10$	$\alpha = 0.4, \beta = 0.05$
V0	**42,686,809**	**33,598,175**	**43,633,497**
	(0.0000)	(0.0000)	(0.0000)
V1	43,891,684	34,427,911	44,880,200
	(0.0282)	(0.0247)	(0.0286)
V2	35,041,939	35,041,939	21,461,941
	(−0.1791)	(0.043)	(−0.5081)

Dividend option/variation of the surplus determination

Table 5.32 Values of the dividend option in Swiss Francs using different surplus determination. The values in brackets denote the relative difference to the basic simulation run

Component	Surplus parameter combination		
Single premium	$\alpha = 0.8, \beta = 0.10$	$\alpha = 0.4, \beta = 0.10$	$\alpha = 0.4, \beta = 0.05$
V0	**89,501,240**	**87,696,664**	**89,404,943**
	(0.0000)	(0.0000)	(0.0000)
V1	89,826,269	87,765,823	89,761,800
	(0.0036)	(0.0008)	(0.0040)
V2	64,261,665	64,261,665	36,636,855
	(−0.2820)	(−0.2672)	(−0.5902)
Regular premium	$\alpha = 0.8, \beta = 0.10$	$\alpha = 0.4, \beta = 0.10$	$\alpha = 0.4, \beta = 0.05$
V0	**893,824,435**	**878,842,480**	**901,393,649**
	(0.0000)	(0.0000)	(0.0000)
V1	896,566,168	880,658,970	904,213,892
	(0.0031)	(0.0021)	(0.0031)
V2	624,782,002	624,782,002	526,247,963
	(−0.3010)	(−0.2891)	(−0.4162)

Shareholder option/variation of the surplus determination

Table 5.33 Values of the shareholder option in Swiss Francs using different surplus determination. The values in brackets denote the relative difference to the basic simulation run

Component	Surplus parameter combination		
Single premium	$\alpha = 0.8, \beta = 0.10$	$\alpha = 0.4, \beta = 0.10$	$\alpha = 0.4, \beta = 0.05$
V0	**18,358,262**	**19,607,117**	**18,048,894**
	(0.0000)	(0.0000)	(0.0000)
V1	18,151,386	19,512,360	17,794,845
	(−0.0113)	(−0.0048)	(−0.0141)
V2	17,279,297	17,279,297	18,977,637
	(−0.0588)	(−0.1187)	(0.0515)
Regular premium	$\alpha = 0.8, \beta = 0.10$	$\alpha = 0.4, \beta = 0.10$	$\alpha = 0.4, \beta = 0.05$
V0	**102,467,939**	**111,707,333**	**101,860,543**
	(0.0000)	(0.0000)	(0.0000)
V1	101,270,633	110,963,272	100,536,575
	(−0.0117)	(−0.0067)	(−0.0130)
V2	105,016,561	105,016,561	116,000,240
	(0.0249)	(−0.0599)	(0.1388)

5.9 Simulating a Mortality Shock

The last aspect of this sensitivity analysis to be considered shall be some sort of mortality shock. Mortality rates are assumed to be deterministic and hence there is no variability in it. What we can basically do to anyhow analyse the influence of mortality rates on the values of the embedded options is to perform a stress test. For this reason, two more simulation runs are performed whereas the yearly mortality rates from the basic simulation run, EKM/F_01_05_II, are multiplied by a factor 1.1 and 0.9 respectively. For both kind of shocks, the mortality rates for the terminal ages $\omega = 116$ (men) and $\omega = 118$ (women) are necessarily still assumed to be equal to one.

Below Tables 5.34–5.36 show the results.

We observe two things. First of all, the impacts of the mortality shock are nearly symmetrically distributed around the reference results of the basic simulation run. Second, a mortality shock of 10%, whether plus or minus, has almost no effect regarding the embedded options. Both arguments can be attributed to the composition of the portfolios that are used for the modelling. The vast majority are endowment insurance which are only poorly affected by this dimension of mortality shocks. If we were considering annuities, for example, the impacts would clearly be more dramatic.

Interest rate guarantee/simulating a mortality shock

Table 5.34 Values of the interest rate guarantee in Swiss Francs simulating a mortality shock. The values in brackets denote the relative difference to the basic simulation run

Component	Surplus parameter combination		
Single premium	$\alpha = 0.8, \beta = 0.10$	$\alpha = 0.4, \beta = 0.10$	$\alpha = 0.4, \beta = 0.05$
Plus 10%	39,980,148	37,142,189	39,772,806
	(0.0122)	(0.0133)	(0.0123)
Best estimate	**39,497,246**	**36,653,299**	**39,288,592**
	(0.0000)	(0.0000)	(0.0000)
Minus 10%	39,022,924	36,173,327	38,813,452
	(−0.012)	(−0.0131)	(−0.0121)
Regular premium	$\alpha = 0.8, \beta = 0.10$	$\alpha = 0.4, \beta = 0.10$	$\alpha = 0.4, \beta = 0.05$
Plus 10%	42,485,247	33,461,433	43,491,373
	(−0.0047)	(−0.0041)	(−0.0033)
Best estimate	**42,686,809**	**33,598,175**	**43,633,497**
	(0.0000)	(0.0000)	(0.0000)
Minus 10%	42,892,995	33,739,160	43,779,320
	(0.0048)	(0.0042)	(0.0033)

Dividend option/simulating a mortality shock

Table 5.35 Values of the dividend option in Swiss Francs simulating a mortality shock. The values in brackets denote the relative difference to the basic simulation run

Component	Surplus parameter combination		
Single premium	$\alpha = 0.8, \beta = 0.10$	$\alpha = 0.4, \beta = 0.10$	$\alpha = 0.4, \beta = 0.05$
Plus 10%	87,563,704	85,720,940	87,436,010
	(−0.0216)	(−0.0225)	(−0.0220)
Best estimate	**89,501,240**	**87,696,664**	**89,404,943**
	(0.0000)	(0.0000)	(0.0000)
Minus 10%	91,655,521	89,897,845	91,599,116
	(0.0241)	(0.0251)	(0.0245)
Regular premium	$\alpha = 0.8, \beta = 0.10$	$\alpha = 0.4, \beta = 0.10$	$\alpha = 0.4, \beta = 0.05$
Plus 10%	879,042,954	864,097,738	886,494,742
	(−0.0165)	(−0.0168)	(−0.0165)
Best estimate	**893,824,435**	**878,842,480**	**901,393,649**
	(0.0000)	(0.0000)	(0.0000)
Minus 10%	908,757,475	893,738,941	916,434,709
	(0.0167)	(0.0170)	(0.0167)

Shareholder option/simulating a mortality shock

Table 5.36 Values of the shareholder option in Swiss Francs simulating a mortality shock. The values in brackets denote the relative difference to the basic simulation run

Component	Surplus parameter combination		
Single premium	$\alpha = 0.8, \beta = 0.10$	$\alpha = 0.4, \beta = 0.10$	$\alpha = 0.4, \beta = 0.05$
Plus 10%	18,179,207	19,404,231	17,862,374
	(−0.0098)	(−0.0103)	(−0.0103)
Best estimate	**18,358,262**	**19,607,117**	**18,048,894**
	(0.0000)	(0.0000)	(0.0000)
Minus 10%	18,538,888	19,812,524	18,239,124
	(0.0098)	(0.0105)	(0.0105)
Regular premium	$\alpha = 0.8, \beta = 0.10$	$\alpha = 0.4, \beta = 0.10$	$\alpha = 0.4, \beta = 0.05$
Plus 10%	101,978,914	111,140,080	101,349,021
	(−0.0048)	(−0.0051)	(−0.0050)
Best estimate	**102,467,939**	**111,707,333**	**101,860,543**
	(0.0000)	(0.0000)	(0.0000)
Minus 10%	102,952,742	112,272,172	102,376,853
	(0.0047)	(0.0051)	(0.0051)

In reality, the probability of a 10% increase or decrease of the yearly mortality rates from 1 year to another is fairly small and so it does not make sense to assume even a bigger shock whether for modelling purposes or not.

With this last sensitivity analysis the empirical analysis ends. There will be an additional section to summarise the results of the sensitivity analysis before I will conclude on this thesis as a whole in the upcoming Chap. 6.

5.10 A Conclusion on the Sensitivity Analysis

The sensitivity analysis performed above gives important information on the functionality of the entire modelling algorithm. By this, the empirical analysis of this thesis ends and so it is the moment to conclude on the sensitivity analysis by summarising the most relevant statements that were deduced therefrom:

- The relationship between the shareholder option and the shareholder margin is almost linear. A lower shareholder margin naturally reduces the value of the shareholder option as well as the value of the interest rate guarantee. But, the joint value of the shareholder option and the dividend option raises as one decreases the shareholder margin.
- From the three different interest rate models the CIR model turns out to be unsuitable for the purpose of this modelling approach as it significantly underestimates the interest rate risk in terms of the low values of the interest rate guarantee. The quantitative differences between the HW and the G2++ model are identified. In terms of the higher total values of all the embedded options, the G2++ is a more conservative model than the HW model. Together with the theoretical background and the description of the calibration process, the user should be able to decide whether to use the HW approach or the G2++ approach for his own purposes.
- The application of various asset allocation strategies demonstrates how sensitive this model responds to this input. Under the given assumptions, the more progressive asset allocation clearly outperforms the realistic and the more conservative asset allocations with respect to the values of the embedded options. For sure, the results were impressive but anyhow have to be interpreted carefully. Nevertheless, this sensitivity analysis enables a reasonable and powerful basis to perform ALM actions.
- A comparison of the end of 2004 and end of 2009 basic simulation runs impressively demonstrates where the limitations of the market consistent valuation methodology are when applying a modelling approach mixing both, market consistent and statutory valuation methodology. According to this, insurers should successively move towards a comprehensive and entirely consistent application of the market consistent methodologies regarding valuation, reserving and, as far as possible, also pricing. This is the corollary to get even more meaningful

values for the embedded options in particular and to attain an entirely consistent risk management in general.

- Using the zero coupon rates as given in the SST framework yields a higher present value of the policy account and thus reflects a more conservative behaviour in the interest rate assumptions but anyhow produces lower total values for the interest rate guarantee and higher total values for the dividend and the shareholder option.
- Assuming premium reduction instead of interest-bearing accumulation as the surplus distribution mechanism reduces the values of the dividend option and the interest rate guarantee due to a change in the surplus payout pattern. On the other hand, the shareholder option is only poorly affected by this changing.
- Interchanging the order of shareholder and policyholder profit has only little impact on the values of the embedded options. The results of the alternatively suggested definition of the policyholder profit are difficult to compare with the default case since both approaches are structurally different. So the redefinition is rather deemed to be a heuristic attempt to possibly improve the default definition and to motivate the reader to create its own customised definition for company specific purposes.
- A mortality shock of either plus or minus 10% has symmetrical, but almost negligible impacts on the portfolios regarding the values of the embedded options.

With this summary this chapter ends. To conclude on this thesis as a whole I will pick up the objectives that are outlined in the first chapter and critically discuss the insights that are gained by the presented modelling framework. Finally there shall be an outlook for further research opportunities on this area.

Chapter 6
Conclusion

This thesis is mainly motivated by the history of the Equitable Life that has shown how threatening embedded options and guarantees can become for the financial stability of an insurance company. The collapse of one of the oldest and most renowned life insurers as a result of its embedded options is, amongst others, one of the reasons that caused a far-reaching change in the local insurance supervisory and general accounting methodology taking place all over the world at the beginning of this century. As a consequence revised solvency frameworks like the SST in Switzerland or Solvency II in the EU request insurers to explicitly allow for embedded options and guarantees in the calculating of their solvency capital. Embedded options and guarantees are also a major concern within the framework of the MCEV, which is rather a valuation framework than a solvency framework.

Following the theoretical specifications of these solvency and valuation frameworks, a numerical model was designed in order to empirically analyse a set of more than 110,000 real existing life insurance policies with regard to the embedded options. The model makes use of a simple balance sheet approach with a stochastic simulation process on the asset side. Depending on this, a deterministic cash flow based process is performed on the liability side. Employing this model, the current market consistent values of selected implicit embedded options are calculated: Interest rate guarantee, dividend option and shareholder option.

The empirical analysis starts by valuing the embedded options under best estimate assumptions, that is to say realistic or, to my mind, most appropriate assumptions. Its results are the answer to the central question I asked in the Objectives: *What value do the considered embedded options and guarantees have, absolutely and relative to the current policy reserves, given a perspective taken from the insurer, the policyholder and the shareholder respectively?*

As for the whole thesis, *current* is meant to be as at the end of 2009.

N. Rüfenacht, *Implicit Embedded Options in Life Insurance Contracts*,
Contributions to Management Science, DOI 10.1007/978-3-7908-2843-6_6,
© Springer-Verlag Berlin Heidelberg 2012

Taking the perspective of the policyholder or the shareholder is equivalent to considering the dividend or the shareholder option respectively. The perspective of the insurer is meant to be considering all of the embedded options as a whole. Below there is Table 6.1 summarising these values based on all the available data under best estimate assumptions with a distribution ratio $\alpha = 0.8$ and a target buffer ratio $\beta = 0.1$, that shall be representative set of surplus parameters. The shareholder margin sh_margin is equal 0.005. The relative values of the embedded options are obtained by dividing by the current policy reserves (including the current surplus reserves) that is equal to the initial asset portfolio A_0.

Table 6.1 Total values in Swiss Francs of the interest rate guarantee, the dividend option and the shareholder option under best estimate assumptions. The underlying data basis consists of 25,707 single premium policies and 85,730 regular premium policies with an initial asset portfolio A_0 of 1,157,664,264 Swiss Francs (single premium) and 1,642,879,688 Swiss Francs (regular premium)

$\alpha = 0.8, \beta = 0.10$, sh_margin $= 0.005$	Single premium	Regular premium
Total interest rate guarantee	39,497,246	42,686,809
Interest rate guarantee per policy	1,536	498
Interest rate guarantee relative to A_0	0.0341	0.0260
Total dividend option	89,501,240	893,824,435
Dividend option per policy	3,482	10,426
Dividend option relative to A_0	0.0773	0.5441
Total shareholder option	18,358,262	102,467,939
Shareholder option per policy	714	1,195
Shareholder option relative to A_0	0.0159	0.0624
Total of all options	147,356,748	1,038,979,183
Total of all options per policy	5,732	12,119
Total of all options relative to A_0	0.1273	0.6324

What do these results tell us? From the perspective of the insurer the values of these embedded options quantify the market consistent liability as at the end of 2009 under best estimate assumptions. It is essential to understand the exact meaning of the values of these options. Since the dividend and the shareholder option are implicitly hedged by definition already, the insurer does not need to set up a reserve for these options today. Its values basically quantify the worth additionally offered to the policyholders and the shareholders due to the currently applied surplus policy.

Different than surpluses the insurer guarantees crediting a fixed minimum interest rate on the savings premiums and this guarantee is neither implicitly nor explicitly hedged in this model. Thus, the value of the interest rate guarantee actually quantifies the amount the insurer should reserve today in order to meet this future obligations.

From the perspective of the policyholder and the shareholder, the values quantify the market consistent value of their additional but non guaranteed benefits as at the end of 2009.

A sensitivity analysis is performed to support the informative value of these results and to consolidate the modelling approach itself. There shall be a focus on the modelling of interest rates due to the sensitivity of interest rates on the embedded options.

The stochastic ZCB prices are modelled as a part of the entire asset modelling process. Therefore, the impacts of three different interest rate models are considered: The CIR model, the HW model and the G2++ model. At the same time, this particular issue gives attention to potential interest rate model risks.

However, the CIR model turns out to be unsuitable for the purpose of this modelling approach as it significantly underestimates the interest rate risk in terms of the low values of the interest rate guarantee.

The quantitative differences between the HW and the G2++ model are identified. In terms of the higher total values of all the embedded options, the G2++ is a more conservative model than the HW model. Together with the theoretical background and the description of the calibration process, the user should be able to decide whether to use the HW or the G2++ approach for his own purposes. A summary on the total option values for the different interest rate models is highlighted in Table 6.2.

Table 6.2 The total option value in Swiss Francs for different interest rate models. The values in brackets denote the relative difference to the model used within the basic simulation run, the G2++ model

$\alpha = 0.8, \beta = 0.10$, sh_margin $= 0.005$	Single premium	Regular premium
Total option value for the CIR model	109,907,929 (−0.2541)	860,577,698 (−0.1717)
Total option value for the HW model	137,484,798 (−0.0670)	1,024,063,971 (−0.0144)
Total option value for the G2++ model	147,356,748 (0.0000)	1,038,979,183 (0.0000)

At the same time different interest rate curves for discounting purposes are analysed. Both, Solvency II and the MCEV calculations, prescribe the use of the swap rate curve. An alternative comes in the form of the zero coupon curve, as it is required for the SST calculations. Using the current zero coupon rates, that are on average approximately 50 basis points below the current swap rates, yields a higher present value of the policy account and thus reflects a more conservative behaviour in the interest rate assumptions but anyhow produces lower total values for the interest rate guarantee and higher total values for the dividend and the shareholder option. The total values of all the embedded options are summarised in Table 6.3.

Currently there is a lively discussion around the SST whether to go on using the zero coupon rates for discounting or to adapt to the international standard Solvency II that uses the swap rates. Potentially these results may support this discussion from a quantitative perspective.

Table 6.3 The total option value in Swiss Francs for different market interest rates. The values in brackets denote the relative difference to the interest rates used within the basic simulation run, the swap rates

$\alpha = 0.8, \beta = 0.10$, sh_margin = 0.005	Single premium	Regular premium
Total option value using zero coupon rates	158,288,684 (0.0742)	1,128,444,234 (0.0861)
Total option value using swap rates	147,356,748 (0.0000)	1,038,979,183 (0.0000)

The most fundamental statement of this thesis is deduced from a comparison over time. Performing best estimate simulation runs as at the end of 2004 and 2009 impressively demonstrates where the limitations of the market consistent valuation methodology are when applying a modelling approach mixing both, market consistent and statutory valuation methodology. According to this, insurers should successively move towards a comprehensive and entirely consistent application of the market consistent methodologies regarding valuation, reserving and, as far as possible, also pricing. This is the corollary to get even more meaningful values for the embedded options in particular and to attain an entirely consistent risk management in general.

After this summary of the major insights I will finally highlight the advantages of the entire modelling approach and critically discuss probable points of improvement.

What are the advantages of this modelling approach? It enables determining the current market consistent values of the interest rate guarantee, the dividend and the shareholder option embedded in either a portfolio of endowment, pure endowment, term life or whole life insurance. The modelling algorithm is formulated in an universally valid manner. Whenever possible, publicly available data is used. Thus, it can be customised for one's personal purposes with respect to, for example, the surplus determination mechanism, the asset modelling process or taking explicit embedded options into account.

Furthermore, the modelling approach opens up opportunities that are not directly associated with the valuation of the embedded options like simple ALM by varying the asset allocation. Indeed, a more detailed consideration of the ALM would clearly request a further customised and more sophisticated modelling algorithm, but this is beyond the scope of this thesis. Besides, consequently market consistent valuation principles are applied. Since these principles have become an absolute standard in the current solvency and valuation frameworks, this modelling approach also sensitises the user to the methodology of market consistency.

What are the points of improvement for this modelling approach? Naturally, a model is always a simplification of the real world that often bears a conflict of goals with respect to the required complexity and the allowed simplification. The model as it is presented arises from the mind of the author in order to meet the objectives of this thesis. And it is further, to a certain degree, governed by the available data basis. Thus, one question to be asked is whether or not the reader accepts this model with all the assumptions that are made. According to this I see the greatest potential to improve the current model by customising it for one's individual purposes.

6.1 Outlook for Further Research

We have seen that the current modelling approach unifies many effective issues that form a powerful valuation framework with a focus on implicit embedded options. Nevertheless, there are still opportunities to improve this modelling approach. I see three interesting tasks for further research.

Within this thesis no explicit embedded options are considered. But conferring Chap. 1, we constitute that this is actually requested by the guidelines of the there introduced frameworks. Thus, as a natural step, the modelling approach should be extended towards a consideration of explicit embedded options. For sure this presupposes a profound statistical analysis of an insurer's data in order to be able to model the behaviour of the policyholders.

The asset modelling process basically defines a "buy-and-hold" strategy without any hedging opportunities. In practice, this approach is of course too simple. Thus, this process could be improved by implementing an appropriate hedging strategy and generally defining an alternative, customised investment strategy for the investment process.

The last issue I mention concerns the different valuation principles like statutory and market consistent. The empirical analysis demonstrates an apparent conflict when applying a modelling approach mixing both, market consistent and statutory valuation methodology. According to this, insurers should successively move towards a comprehensive and entirely consistent application of the market consistent methodologies regarding valuation, reserving and, as far as possible, also pricing. This is the corollary to get even more meaningful values for the embedded options in particular and to attain an entirely consistent risk management in general. Indeed this is a strong postulation, but it is deemed as an attempt to motivate researchers as well as practitioners to further proceed the development of this area of research.

Appendix A
Calibration Data Basis

This chapter provides the data basis used to calibrate the stochastic simulation processes as presented in Chap. 3. For the data basis as at the end of 2009 see Sect. A.1 (Figs. A.1, A.2, and Table A.1), for the data basis including the calibration results as at the end of 2004 see Sect. A.2 (Figs. A.3, A.4, and Tables A.2, A.3).

A.1 Calibration Data Basis as at the End of 2009

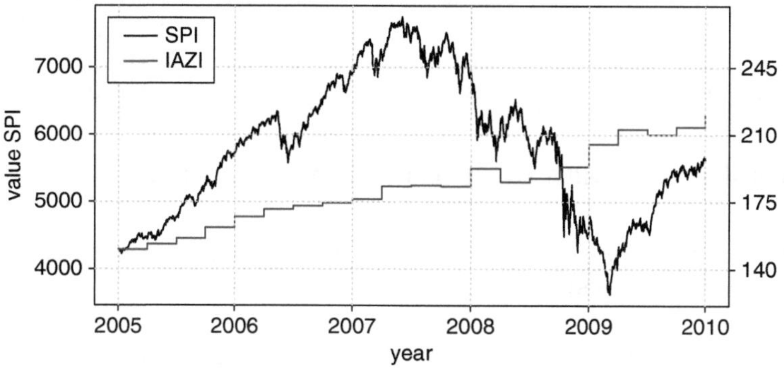

Fig. A.1 Historical chart of the SPI and the IAZI index from 2005 to 2009 (Source: Datastream)

N. Rüfenacht, *Implicit Embedded Options in Life Insurance Contracts*, Contributions to Management Science, DOI 10.1007/978-3-7908-2843-6,
© Springer-Verlag Berlin Heidelberg 2012

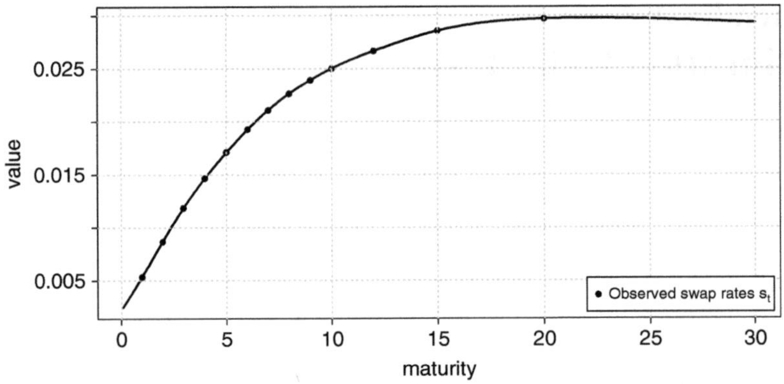

Fig. A.2 (Current) Swap rate curve (CHF) as at 31.12.2009. The values between the observed swap rates s_t were interpolated performing the cubic spline procedure `spline(...)` in **R** (Source: Datastream)

Table A.1 ATM European implied swaption volatility quotes (CHF) in % as at 31.12.2009 (Source: Datastream)

	Swap length in years									
Option maturity	1	2	3	4	5	6	7	8	9	10
1	55.4	42.8	39.7	37.8	35.3	33.1	31 .0	29.6	28.3	27.3
2	45.6	37.4	34.6	32.6	30.8	29.3	28.1	27.1	26.2	25.5
3	37.2	31.2	29.4	28.1	27.1	26.2	25.4	24.7	24.2	23.7
4	31.9	26.9	25.8	25.0	24.4	23.9	23.4	23.0	22.6	22.3
5	27.3	24.0	23.4	23.0	22.7	22.3	22.0	21.7	21.5	21.3
7	22.6	20.9	20.7	20.6	20.6	20.4	20.3	20.2	20.1	20.1
10	19.3	18.7	18.8	18.9	19.1	19.2	19.3	19.5	19.6	19.8

A.2 Calibration Data Basis as at the End of 2004

Remark: The share price process $S(t)$ and the property price process $P(t)$ are defined by the following SDE.

$$dS(t) = \alpha S(t)dt + \sigma S(t)dW^1(t), \quad S_0 = 1,$$

$$dP(t) = \beta P(t)dt + \eta P(t)dW^2(t), \quad P_0 = 1,$$

where (W^1, W^2) is a two-dimensional Brownian motion with instantaneous correlation ρ as from

$$dW^1(t)dW^2(t) = \rho dt.$$

To get further information on it confer Sect. 3.3.

Table A.2 Monthly recorded swap rates (CHF) s_t for 2009. ZCB prices can be derived from these rates according to $P(0,t) = \exp(-t \cdot s_t)$ (Source: Datastream)

Time		Swap length in years												
t		1	2	3	4	5	6	7	8	9	10	12	15	20
31.12.2009	s_t	0.0054	0.0087	0.0119	0.0146	0.0172	0.0192	0.0210	0.0226	0.0239	0.0250	0.0267	0.0286	0.0297
	$P(0,t)$	0.9947	0.9828	0.9649	0.9432	0.9178	0.8910	0.8631	0.8347	0.8065	0.7788	0.7256	0.6509	0.5527
30.11.2009	s_t	0.0054	0.0079	0.0112	0.0141	0.0165	0.0185	0.0203	0.0217	0.0230	0.0241	0.0259	0.0278	0.0290
	$P(0,t)$	0.9947	0.9844	0.9670	0.9454	0.9210	0.8949	0.8678	0.8404	0.8132	0.7862	0.7330	0.6595	0.5605
30.10.2009	s_t	0.0054	0.0087	0.0119	0.0149	0.0174	0.0195	0.0213	0.0227	0.0240	0.0251	0.0269	0.0289	0.0301
	$P(0,t)$	0.9947	0.9828	0.9649	0.9423	0.9168	0.8897	0.8617	0.8343	0.8056	0.7780	0.7239	0.6482	0.5477
30.09.2009	s_t	0.0054	0.0079	0.0112	0.0142	0.0167	0.0189	0.0208	0.0223	0.0236	0.0248	0.0267	0.0287	0.0301
	$P(0,t)$	0.9947	0.9843	0.9669	0.9450	0.9198	0.8928	0.8648	0.8364	0.8089	0.7803	0.7263	0.6499	0.5479
31.08.2009	s_t	0.0054	0.0074	0.0107	0.0137	0.0164	0.0186	0.0206	0.0223	0.0236	0.0248	0.0267	0.0287	0.0300
	$P(0,t)$	0.9947	0.9853	0.9686	0.9469	0.9215	0.8944	0.8657	0.8367	0.8089	0.7806	0.7263	0.6499	0.5488
31.07.2009	s_t	0.0075	0.0109	0.0140	0.0167	0.0191	0.0210	0.0225	0.0239	0.0249	0.0267	0.0285	0.0299	0.0306
	$P(0,t)$	0.9925	0.9784	0.9589	0.9354	0.9091	0.8819	0.8543	0.8263	0.7992	0.7661	0.7103	0.6388	0.5423
30.06.2009	s_t	0.0054	0.0082	0.0120	0.0156	0.0189	0.0213	0.0235	0.0250	0.0263	0.0273	0.0289	0.0307	0.0319
	$P(0,t)$	0.9947	0.9837	0.9648	0.9394	0.9098	0.8799	0.8486	0.8185	0.7892	0.7613	0.7067	0.6312	0.5280
29.05.2009	s_t	0.0054	0.0077	0.0113	0.0151	0.0186	0.0215	0.0236	0.0253	0.0269	0.0280	0.0296	0.0312	0.0318
	$P(0,t)$	0.9947	0.9848	0.9668	0.9413	0.9111	0.8790	0.8477	0.8166	0.7853	0.7558	0.7008	0.6263	0.5291
30.04.2009	s_t	0.0054	0.0082	0.0116	0.0147	0.0174	0.0196	0.0213	0.0227	0.0239	0.0249	0.0263	0.0276	0.0279
	$P(0,t)$	0.9947	0.9837	0.9659	0.9431	0.9168	0.8893	0.8615	0.8337	0.8065	0.7800	0.7295	0.6607	0.5726
30.03.2009	s_t	0.0054	0.0077	0.0111	0.0144	0.0173	0.0196	0.0214	0.0228	0.0240	0.0250	0.0265	0.0276	0.0274
	$P(0,t)$	0.9947	0.9847	0.9673	0.9441	0.9171	0.8891	0.8609	0.8334	0.8061	0.7792	0.7276	0.6607	0.5783
27.02.2009	s_t	0.0068	0.0092	0.0126	0.0157	0.0184	0.0205	0.0222	0.0236	0.0247	0.0257	0.0271	0.0281	0.0276
	$P(0,t)$	0.9933	0.9818	0.9630	0.9392	0.9123	0.8845	0.8564	0.8283	0.8007	0.7738	0.7228	0.6558	0.5758
30.01.2009	s_t	0.0067	0.0089	0.0121	0.0151	0.0176	0.0197	0.0213	0.0227	0.0238	0.0248	0.0262	0.0275	0.0274
	$P(0,t)$	0.9933	0.9823	0.9643	0.9415	0.9158	0.8888	0.8615	0.8343	0.8072	0.7808	0.7300	0.6625	0.5787

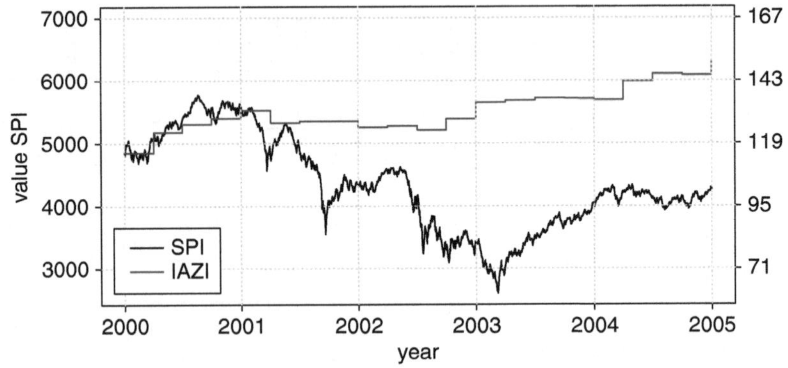

Fig. A.3 Historical chart of the SPI and the IAZI index from 2000 to 2004 (Source: Datastream)

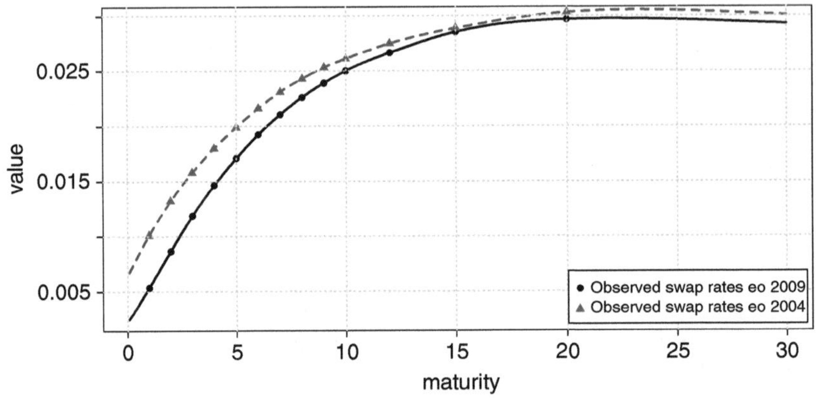

Fig. A.4 Comparison of the end of 2004 (eo 2004) and the end of 2009 (eo 2009) swap rate curve (CHF). The values between the observed swap rates were interpolated performing the cubic spline procedure `spline(...)` in **R** (Source: Datastream)

Table A.3 Estimated parameters for the share price process and the property price process to calibrate as at the end of 2004 (Eo 2004) and 2009 (Eo 2009)

Parameter	Eo 2004	Eo 2009
α	−0.0034	0.0736
σ	0.1983	0.1963
β	0.05357	0.0781
η	0.0498	0.0444
ρ	0.0024	−0.1014

For the purpose of the sensitivity analysis the G2++ interest rate model is calibrated to swaption implied volatilities as at the end of 2004. Since the earliest data for interest rate swaps (CHF) that could be found in Datastream were as at 28.02.2005, also the swaption implied volatilities were taken as at this date as they were as at the end of 2004. See Table A.4 for the ATM European implied swaption volatility quotes (CHF) as at 28.02.2005.

Table A.4 ATM European implied swaption volatility quotes (CHF) in % as at 28.02.2005 (Source: Datastream)

		Swap length in years									
		1	2	3	4	5	6	7	8	9	10
Option maturity	1	43.0	36.3	31.7	28.4	25.7	23.5	21.7	20.3	19.0	17.9
	2	35.3	31.0	27.7	25.0	23.0	21.4	20.1	19.0	18.0	17.3
	3	29.3	26.5	24.1	22.3	20.9	19.6	18.5	17.7	17.0	16.4
	4	25.5	23.2	21.5	20.1	18.9	18.0	17.3	16.6	16.1	15.6
	5	22.4	20.7	19.5	18.4	17.7	16.9	16.3	15.8	15.4	15.1
	7	18.6	17.7	16.9	16.3	15.8	15.3	14.9	14.6	14.3	14.1
	10	15.8	15.5	15.1	14.8	14.4	14.1	13.9	13.8	13.7	13.7

Based on these swaption implied volatilities the estimated parameters to calibrate the G2++ model as at the end of 2004 are as shown in Table A.5.

Table A.5 Estimated parameters for the G2++ model to calibrate as at the end of 2004 (eo 2004) and 2009 (eo 2009). Ψ denotes the mean squared error term as a measure for the goodness of fit as defined in Eq. 3.67

Parameter	Eo 2004	Eo 2009
a	3.8202	2.3660
b	0.0727	0.0327
σ	0.0641	0.0166
η	0.0081	0.0089
ρ	−0.6370	−0.9462
Ψ	0.00072	0.00225

Remark: Within the G2++ model the time t price for a ZCB with maturity T, $P(t,T)$, is calculated according to

$$P(t,T) = \frac{P^M(0,T)}{P^M(0,t)} \exp\{\mathcal{A}(t,T)\},$$

where

$$\begin{aligned}\mathcal{A}(t,T) &= \frac{1}{2}[V(t,T) - V(0,T) + V(0,t)] \\ &\quad - \frac{1-e^{-a(T-t)}}{a} x(t) - \frac{1-e^{-b(T-t)}}{b} y(t),\end{aligned}$$

$$
\begin{aligned}
V(t,T) = {} & \frac{\sigma^2}{a^2}\left[T - t + \frac{2}{a}e^{-a(T-t)} - \frac{1}{2a}e^{-2a(T-t)} - \frac{3}{2a}\right] \\
& + \frac{\eta^2}{b^2}\left[T - t + \frac{2}{b}e^{-b(T-t)} - \frac{1}{2b}e^{-2b(T-t)} - \frac{3}{2b}\right] \\
& + 2\rho\frac{\sigma\eta}{ab}\left[T - t + \frac{e^{-a(T-t)} - 1}{a} + \frac{e^{-b(T-t)} - 1}{b} - \frac{e^{-(a+b)(T-t)} - 1}{a+b}\right],
\end{aligned}
$$

and P^M denoting the market ZCB price.

Appendix B
Additional Information on the Basic Simulation Run

Within this chapter the reader finds additional information on the behaviour of the simulation algorithm. Based on the basic simulation run as performed in Sect. 5.1, Tables B.1 and B.2 show the variation coefficient for the interest rate guarantee and the dividend option respectively. Find the definition of the variation coefficient below.

Definition 11 (variation coefficient (vc)). Let $\hat{\sigma}_x$ and $\hat{\mu}_x$ be the standard deviation and the mean of a set of samples. Then the variation coefficient (vc) is defined as follows.

$$vc = \frac{\hat{\sigma}_x}{\hat{\mu}_x} \tag{B.1}$$

Regarding the available data there are two categories that allow for a terminal bonus, endowment and pure endowment insurance for regular premiums. Within the numerical analysis presented in Chap. 5, only the dividend option was considered as a whole. Therefore, Table B.3 reports the values of this terminal bonus, that is a part of the total dividend option, based on the basic simulation run.

Finally there is Table B.4 presenting the estimated default probabilities, $\hat{E}[B^{ab}(T^+) < 0]$, again for the basic simulation run. Hence, for a particular simulation a default has occurred if the terminal value of the buffer B is negative.

$$\text{Default} \Leftrightarrow B^{ab}(T^+) < 0 \tag{B.2}$$

Note that the presented default probabilities are mean values for particular categories, based on the estimated default probabilities of each portfolio within a category.

N. Rüfenacht, *Implicit Embedded Options in Life Insurance Contracts*, Contributions to Management Science, DOI 10.1007/978-3-7908-2843-6,
© Springer-Verlag Berlin Heidelberg 2012

Table B.1 Variation coefficient for the interest rate guarantee based on 5,000 basic simulation runs as outlined in Sect. 5.1

Category	Parameter combination		
Single premium	$\alpha = 0.8, \beta = 0.10$	$\alpha = 0.4, \beta = 0.10$	$\alpha = 0.4, \beta = 0.05$
Endowment	2.5133	2.7367	2.5264
Whole life	4.5469	5.2538	4.3181
Pure endowment	1.7540	1.8150	1.7543
Regular premium	$\alpha = 0.8, \beta = 0.10$	$\alpha = 0.4, \beta = 0.10$	$\alpha = 0.4, \beta = 0.05$
Endowment	2.3190	2.4979	2.2543
Whole life	3.1239	3.4082	3.0884
Term life	2.3385	2.9470	2.7766
Pure endowment	1.9610	2.0735	1.8636

Table B.2 Variation coefficient for the dividend option based on 5,000 basic simulation runs as outlined in Sect. 5.1

Category	Parameter combination		
Single premium	$\alpha = 0.8, \beta = 0.10$	$\alpha = 0.4, \beta = 0.10$	$\alpha = 0.4, \beta = 0.05$
Endowment	1.0861	1.0732	1.0665
Whole life	0.3241	0.3538	0.3377
Pure endowment	1.4923	1.4867	1.4595
Regular premium	$\alpha = 0.8, \beta = 0.10$	$\alpha = 0.4, \beta = 0.10$	$\alpha = 0.4, \beta = 0.05$
Endowment	0.5299	0.5467	0.5228
Whole life	0.6187	0.7274	0.6876
Term life	0.0584	0.0604	0.0611
Pure endowment	0.8229	0.8193	0.7953

Table B.3 Values of the terminal bonus denoted in CHF based on 5,000 basic simulation runs. The values in brackets denote the values relative to the total amount of the dividend option

Category	Parameter combination		
Regular premium	$\alpha = 0.8, \beta = 0.10$	$\alpha = 0.4, \beta = 0.10$	$\alpha = 0.4, \beta = 0.05$
Endowment	127,322,261	192,985,212	136,737,414
	(0.1543)	(0.2387)	(0.1646)
Pure endowment	1,424,755	1,866,920	1,313,964
	(0.3697)	(0.5055)	(0.3309)

Table B.4 Estimated mean default probabilities based on 5,000 basic simulation runs as outlined in Sect. 5.1

Category	Parameter combination		
Single premium	$\alpha = 0.8, \beta = 0.10$	$\alpha = 0.4, \beta = 0.10$	$\alpha = 0.4, \beta = 0.05$
Endowment	0.6171	0.4427	0.5579
Whole life	0.4917	0.3192	0.3925
Pure endowment	0.7575	0.6096	0.7325
Regular premium	$\alpha = 0.8, \beta = 0.10$	$\alpha = 0.4, \beta = 0.10$	$\alpha = 0.4, \beta = 0.05$
Endowment	0.5752	0.4339	0.4974
Whole life	0.4279	0.3137	0.3640
Term life	0.5559	0.3004	0.3306
Pure endowment	0.6652	0.6042	0.6588

References

Amemiya T (1973) Regression analysis when the dependent variable is truncated normal. Econometrica 41:997–1016

AVO (2009) Verordnung vom 9. November 2005 über die Beaufsichtigung von privaten Versicherungsunternehmen (Aufsichtsverordnung, AVO), Stand am 1. Januar 2009, SR-Nummer: 961.011. http://www.admin.ch/ch/d/sr/9/961.011.de.pdf. Last checked 1 June 2011

Barsky RB (1989) Why don't the prices of stocks and bonds move together? Am econ rev 79(5):1132–1145

Brigo D, Mercurio F (2006) Interest rate models – Theory and practice. With smile, inflation and credit, 2nd edn. Springer, New York

Briys E, de Varenne F (1997) On the risk of life insurance liabilities: debunking some common pitfalls. J Risk Insur 64(4):673–694

CEIOPS (2010a) Committee of European insurance and occupational pensions supervisors CEIOPS (2010): QIS5 technical specifications. https://www.ceiops.eu/fileadmin/tx_dam/files/consultations/QIS/QIS5/QIS5-technical_specifications_20100706.pdf. Last checked 1 June 2011

CEIOPS (2010b) Committee of European insurance and occupational pensions supervisors CEIOPS (2010): annexes to the QIS5 technical specifications. https://www.ceiops.eu/fileadmin/tx_dam/files/consultations/QIS/QIS5/Annexes-to-QIS5-technical_specifications_20100706.pdf. Last checked 1 June 2011

CFO Forum (2009a) CFO Forum (2009): market consistent embedded value principles. http://www.cfoforum.nl/downloads/MCEV_Principles_and_Guidance_October_2009.pdf. Last checked 1 June 2011

CFO Forum (2009b) CFO Forum (2009): market consistent embedded value basis for conclusions. http://www.cfoforum.nl/downloads/CFO_Forum_MCEV_Basis_for_Conclusions.pdf. Last checked 1 June 2011

Chadburn RG (1998) Controlling solvency and maximizing policyholdersÕ returns: a comparison of management strategies for accumulating with-profits long-term insurance business. Actuarial Research Paper No. 115, Department of Actuarial Science and Statistics, City University

Cox JC, Ingersoll JE, Ross SA (1985) A theory of the term structure of interest rates. Econometrica 53(1985):385–407

Cummins JD, Miltersen KR, Persson S-A (2004) International comparison of interest rate guarantees in life insurance. Conference proceedings ASTIN 2004, Bergen

EC 138 (2009) Directive 2009/138/EC of the European parliament and of the council of 25 November 2009 on the taking-up and pursuit of the business of insurance and reinsurance (solvency II). http://register.consilium.europa.eu/pdf/en/09/st03/st03643-re06.en09.pdf. Last checked 1 June 2011

N. Rüfenacht, *Implicit Embedded Options in Life Insurance Contracts*,
Contributions to Management Science, DOI 10.1007/978-3-7908-2843-6,

© Springer-Verlag Berlin Heidelberg 2012

FINMA (2008) Eidgenšssische Finanzmarktaufsicht FINMA, Rundschreiben 2008/18 Anlagerichtlinien Versicherer, Stand am 20. November 2008, Referenz: FINMA-RS 08/18 "Anlagerichtlinien Versicherer". http://www.finma.ch/d/regulierung/Documents/finma-rs-2008-18.pdf. Last checked 1 June 2011

FOPI (2004a) Swiss federal office of private insurance FOPI (2004): white paper of the Swiss solvency test. http://www.finma.ch/archiv/bpv/download/e/WhitePaperSST_en.pdf. Last checked 1 June 2011

FOPI (2004b) Swiss federal office of private insurance FOPI (2004): technical document on the Swiss solvency test. http://www.finma.ch/archiv/bpv/download/e/SST_techDok_061002_E_wo_Li_20070118.pdf. Last checked 1 June 2011

FSA (2001) Financial services authority (2001) Glick report – Joint opinion of Ian Glick QC and Richard Snowden for the financial services authority. May 2001. http://www.fsa.gov.uk/pubs/other/Equitable_Final_Glick_QC.pdf. Last checked 1 June 2011

Gatzert N (2009) Implicit options in life insurance: an overview. Zeitschrift für die gesamte Versicherungswissenschaft 98:141–164

Gerber HU (1997) Life insurance mathematics, 3rd edn. Springer, Berlin

Grosen A, Jørgensen PL (2000) Fair valuation of life insurance liabilities: the impact of interest rate guarantees, surrender pptions and bonus policies. Insur Math Econ 26:37–57

Grosen A, Jørgensen PL (2002a) Life insurance liabilities at market value: an analysis of insolvency risk, bonus policy, and regulatory intervention rules in a barrier option framework. J Risk Insur 69(1):63–91

Grosen A, Jørgensen PL (2002b) The bonus-crediting mechanism of Danish pension and life insurance companies. J Pension Econ Finance 1:249–268. Cambridge University Press

Haberman S, Ballotta L, Wang N (2003) Modelling and valuation of guarantees in with-profit and unitised with-profit life insurance contracts. Actuarial Research Paper No. 146, Department of Actuarial Science and Statistics, Cass Business School London

Hansen M, Miltersen KR (2002) Minimum rate of return guarantees: the Danish case. Scand Actuarial J 2002(4):280–318

Harrison JM, Kreps DM (1979) Martingales and arbitrage in multiperiod securities markets. J Econ Theory 20:381–408

Harrison JM, Pliska SR (1981) Martingales and stochastic integrals in the theory of continuous trading. Stoch Proc Appl 11:215–260

House of Lords (2000) Judgments – Equitable life assurance society v. Hyman on 20 July 2000. http://www.publications.parliament.uk/pa/ld199900/ldjudgmt/jd000720/equite-1.htm. Last checked 1 June 2011

Hull JC (2006) Options, futures and other derivatives, 6th edn. Pearson, Upper Saddle River

Hull JC, White A (1990) Pricing interest rate derivative securities. Rev Financ Stud 3(4):573–592

Hull JC, White A (1994) Branching out. RISK July:34–37

Jensen B, Jørgensen PL, Grosen A (2001) A finite difference approach to the valuation of path dependent life insurance liabilities. Geneva Pap Risk Ins Theory 26:57–84

Keller P, Gisler A, Wüthrich MV (2011) Zur Diskontierung der Versicherungsverpflichtungen im SST. Advisory opinion published by the SAA. http://www.actuaries.ch/50_richtlinien/12_stellungnahmen.htm. Last checked 1 June 2011

Kling A, Richter A, Russ J (2007) The impact of surplus distribution on the risk exposure of with profit life insurance policies including interest rate guarantees. J Risk Insur 74(3):571–589

Menner M, Ilgmann C (2011) Negative nominal interest rates: history and current proposals. CAWM Discussion Papers No. 43 from Center of Applied Economic Research Münster (CAWM), University of Münster

Merton RC (1977) An analytic derivation of the cost of deposit insurance and loan guarantees. J Bank Financ 1:3–11

Merton RC (1990) Continuous time finance. Blackwell, Cambridge

Needleman PD, Roff TA (1995) Asset shares and their use in the financial management of a with-profits fund. Br Actuarial J 1:603–688

Privault N (2008) An elementary introduction to stochastic interest rate modeling. Advanced series of statistical science & applied probability, vol 12. World Scientific, Singapore

Schrager D, Pelsser AAJ (2006) Pricing swaptions and coupon bond options in affine term structure models. Math Financ 16:673–694

Shreve SE (2004) Stochastic calculus for finance II, continuous-time models. Springer, New York

Stivers C, Sun L (2002) Stock market uncertainty and the relation between stock and bond returns. Working Paper 2002–3, Federal Reserve Bank of Atlanta

Swiss Association of Actuaries SAA (2006) Guideline on the market-consistent valuation and modelling of options and guarantees under the Swiss Solvency Test (SST). http://www.actuaries.ch/de/50_richtlinien/01_richtlinien.htm. Last checked 1 June 2011

Swiss Reinsurance Company (2006) Sigma, No 4/2006, solvency II: an integrated risk approach for European insurers. http://media.swissre.com/documents/sigma4_2006_en.pdf. Last checked 1 June 2011

The R Development Core Team (2010) R: a language and environment for statistical computing. Reference index, Version 2.12.1. http://cran.r-project.org/doc/manuals/fullrefman.pdf. Last checked 1 June 2011

Tillinghast-Towers Perrin (2001) Asset share survey 2000. London

VAG (2009) Bundesgesetz vom 17. Dezember 2004 betreffend die Aufsicht über Versicherungsunternehmen (Versicherungsaufsichtsgesetz, VAG), Stand am 1. Januar 2009. SR-Nummer: 961.01. http://www.admin.ch/ch/d/sr/9/961.01.de.pdf. Last checked 1 June 2011

van Haastrecht A (2007) Hybrid equity-interest rate models with focus on efficient simulation and calibration methods. Master thesis, faculty of economics and business administration, vrije universiteit Amsterdam

Vasicek O (1977) An equilibrium characterization of the term structure. J Financ Econ 5:177–188

VVG (2009) Bundesgesetz vom 2. April 1908 über den Versicherungsvertrag (Versicherungsvertragsgesetz, VVG), Stand am 1. Juli 2009, SR-Nummer: 221.229.1. http://www.admin.ch/ch/d/sr/2/221.229.1.de.pdf. Last checked 1 June 2011

Index

N. Rüfenacht, *Implicit Embedded Options in Life Insurance Contracts*,
Contributions to Management Science, DOI 10.1007/978-3-7908-2843-6,
© Springer-Verlag Berlin Heidelberg 2012

I

L

M

P

Q

R

S

T

V

W

Z

Curriculum Vitae

Nils Rüfenacht von Meikirch (Bern) wurde am 28. Juli 1981 in Luzern als Sohn von Marlys Rüfenacht und Thomas Zaugg geboren.

Seine obligatorische Schulzeit hat er an den Schulen in Birsfelden durchlaufen und im Dezember 2000 am Gymnasium Muttenz seine Matura (Typus E) abgelegt.

Anschliessend studierte er von Oktober 2001 - September 2006 Versicherungswissenschaften an der Universität Basel mit dem Nebenfach Mathematik und den Wahlfächern Informatik und Betriebswirtschaftslehre, ehe er im Frühjahr 2007 am Institut für Finanzmarkttheorie der Universität Basel mit der Ausarbeitung seiner Dissertation begann. Die Dissertation entstand unter Anleitung von Herrn Prof. Dr. Heinz Zimmermann und Herrn Prof. Dr. Herbert Lüthy.

Neben seiner Forschungstätigkeit hat Nils Rüfenacht Praktika bei Herrn Dr. Bruno Koller, den Basler Versicherungen sowie den PAX Versicherungen absolviert und arbeitet seit Oktober 2007 als Aktuar für die Helvetia Versicherungen im Bereich Mathematik Leben.

Davos, den 17. Februar 2011,
Nils Rüfenacht

N. Rüfenacht, *Implicit Embedded Options in Life Insurance Contracts*,
Contributions to Management Science, DOI 10.1007/978-3-7908-2843-6,
© Springer-Verlag Berlin Heidelberg 2012

Curriculum Vitae

FSC
www.fsc.org
MIX
Papier aus verantwortungsvollen Quellen
Paper from responsible sources
FSC® C105338

If you have any concerns about our products,
you can contact us on
ProductSafety@springernature.com

In case Publisher is established outside the EU,
the EU authorized representative is:
Springer Nature Customer Service Center GmbH
Europaplatz 3, 69115 Heidelberg, Germany

Printed by Libri Plureos GmbH
in Hamburg, Germany